WILLIAM JAMES BURROUGHS

Does the Weather Really Matter?

the social implications of climate change

CAMBRIDGE
UNIVERSITY PRESS

CAMBRIDGE UNIVERSITY PRESS
Cambridge, New York, Melbourne, Madrid, Cape Town, Singapore, São Paulo

Cambridge University Press
The Edinburgh Building, Cambridge CB2 2RU, UK

Published in the United States of America by Cambridge University Press, New York

www.cambridge.org
Information on this title: www.cambridge.org/9780521561266

First published 1997
This digitally printed first paperback version 2005

A catalogue record for this publication is available from the British Library

Library of Congress Cataloguing in Publication data

Burroughs, William James.
    Does the weather really matter? : the social implications of climate change / William
James Burroughs.
        p.   cm.
    Includes bibliographical references and index.
    ISBN 0 521 56126 4 (hb)
    1. Climatic changes – Economic aspects.   2. Climatic changes – Political aspects.
    I. Title.
QC981.8.C5B87   1997
551.6–dc21   97-492 CIP

ISBN-13 978-0-521-56126-6 hardback
ISBN-10 0-521-56126-4 hardback

ISBN-13 978-0-521-01744-2 paperback
ISBN-10 0-521-01744-0 paperback

Does the Weather Really Matter?

Contents

Preface

'What a silly question!' I hear you say. Any fool knows the weather matters. So why pose the question? The answer lies in the weasel-word 'really'. The whole issue of the potential impact of climatic change, whether natural, or as a result of human activities, depends on how sensitive our economic and social structures are to such changes. Only by asking direct questions about what has been the real impact in the past and how much future developments are likely to take us into new territory can we assess whether the various options for action are worth the effort. This also takes us into difficult areas associated with our ability to forecast and how societies respond to both unexpected changes and to apparently believable forecasts. All these matters have been the subject of an immense amount of expert analysis: UN-sponsored programmes have crawled over issues and drawn on the expertise of a vast number of specialists in the field of meteorology, climate change and its impact on economic and social systems; environmental movements have pressed vigorously to get action on alleviating the worst predictions before they become reality; and leading politicians have nailed their colours to the climatic-change mast.

This leads to a second question: with so much comment and analysis around, why produce another book? The reasons are fourfold. First, there is the issue of perspective. Much of the work, while taking a lengthy view of the future, is inclined to look at only a limited part of the historical evidence of the consequences and nature of past changes. Although there are good reasons for concentrating on both recent events and reliable measurements of how the climate has changed, there is a risk of losing sight of how we have adapted in the past and may adapt in the future. So making certain, we wring as much as possible out of the lessons of the past to get the best possible sense of historical perspective. It pays to take this long view. As Chairman Mao is reputed to have said, when asked what the impact of the French Revolution on history was, 'it is too soon to say.'

The second issue is that of accessibility. The major international ana-
lytical work (e.g. the four volumes produced by the International Panel of
Climate Change, IPCC) is monumental,[1] but it is not readily digested by
the average person. The provision of executive summaries helps, but this
serves principally to present the consensus view and, in so doing, does
not provide the general reader with the more exhilarating flavours of the
intense debate surrounding the questions of the reality of climatic change,
the reliability of forecasts and the potential implications of change. Given
that the handling of these issues by the media often amplifies the more
strident parts of the debate, there is a continuing need to make the arguments as accessible as possible.

This leads naturally to the third point: the question of balance. Some
authors have chosen to simplify the issues which are presented with
admirable balance in the IPCC reports, by taking a partial view of the
arguments. This can give the impression either that future developments
will follow certain paths and hence that immediate massive action is now
essential or that the whole matter is a storm in a teacup. The interests of
all of us are not well served by any such rush to judgement. What is at
stake both in the potential costs of action or inaction and the benefits
flowing from the correct choices is far too great to allow the issues to be
oversimplified. It is vital we have a balanced view of how much we know
and how much we can rely on this knowledge to make politically difficult
decisions about the future allocation of resources to confront the national
and international challenges of climatic change. The political backlash that
has developed in recent years, notably in the USA, against the more
extreme claims of the climatic change community is a good example of
this issue. Unless the arguments are credible to the electorate and their
chosen representatives they will have no impact.

This does not mean I am intent on using uncertainty to avoid making
decisions by ducking behind the smokescreen of needing to do more
research. Whether we like it or not, decisions are being taken now which
will affect our capacity to adapt to whatever climatic changes are in store.
So we may regret having failed to exploit information we already have at
our disposal. Facing up to the need to act now, while recognising that
research will continue that will provide additional information for making
better decisions in the future, is all part of this balance. So making the
most effective use of current knowledge to identify how we can minimise
subsequent scientific, economic and social ramifications of climatic change

needs continual updating. Without this analysis we are not likely to convert the sceptics in business and politics to the need for action.

Finally, there is a need for a sense of realism. The pursuit of an ideal of, say, having a stable climate which is subject only to natural variability, whatever that might be, sounds fine in principle. In practice, however, all human activities will have some impact on the environment, and so the best we can hope to do is minimise the harm we will do. Likewise, building an impressive argument for action on the basis that the impact of predicted global warming over the coming 200 years will be immense is unlikely to cut the mustard with politicians intent on reducing the budget deficit and getting re-elected. So, while we must confront the implications of the attitude exemplified by Maynard Keynes's observation that 'in the long run we are all dead', there is no point in ignoring the political reality that longer term issues rarely intrude into decision-making unless they can be translated into more immediate concerns. This means placing emphasis on options which both appear to contribute to the longer term aims and also make good economic sense now. At the same time, the dangers of over-reacting to isolated extreme events, which appear to confirm certain predictions, is equally relevant, especially where successive events seem to ring contradictory alarm bells.

Notes

1 IPCC (1990), (1992), (1994), (1995).

1

Introduction

'Why, sometimes I've believed as many as
six impossible things before breakfast.'
The White Queen, in *Through the Looking-Glass*, Chapter 5[1]

Dotted around the world are many examples of abandoned cities, monuments and settlements which appear to provide stark evidence of the part fluctuations in the climate can play in human activities. From the jungles of Guatemala to the shores of western Greenland fjords there are the ruins of societies which seem to have been overwhelmed by the effects of prolonged adverse weather. The same story of changing patterns of temperature or rainfall overwhelming day-to-day life can be seen in the Scottish uplands or the deserts of Arizona. The remains of past thriving communities appear to provide mute testimony to the part sustained changes in the weather have played in history.

At a time of growing concern about the impact of human activities on the climate these stark reminders of the impact of natural changes in the climate are seen as potent warnings. But are they relevant? The important questions are 'How big a part did climate change play in both success and failure?' and 'Is the evidence of the past relevant to analysing current events?' Only by addressing these questions can we decide whether examples from the past can be applied to current efforts to predict the consequences of future climatic change.

This process is not easy. The subject of the historical impact of climatic change stirs strong emotions. There is a range between those who argue that fluctuations in the climate were wholly inconsequential in the course of history to those who believe they are an unrecognised mainspring which has controlled the outcome of many events. As with so many fiercely debated issues, reality lies somewhere between these extremes. What is more difficult to establish is precisely where on the spectrum of impact particular events lie. So the objective of this book will be to present examples of where the impact of extreme weather or more sustained shifts of the climate have had measurable consequences and to explore how these have combined with other factors to influence social, political and economic development. Then we will consider just how important predicted changes in the climate are likely to be in comparison with other factors, such as resource constraints or population growth.

Concentrating on the big issues is essential if we are to discover which aspects of our changing weather and climate matter. There is plenty of econometric analysis which shows how changes in, say, temperature or rainfall lead to changes in demand for weather-sensitive goods and services, or alter the supply of other goods and services. Not surprisingly these changes in demand or supply can then be linked to shifts in prices with obvious consequences for producers and consumers. This provides fertile ground for economic analysis of how the market forces exploit the consequent shifts in prices. While this is an important subject on which much has been written,[2] it rarely achieves the level of being a really big issue. What we are looking for here is the instances of how the weather can conspire to have a more permanent effect on the course of events.

This may look like an artificial distinction, and to a certain extent it is. But, it is important to discover which changes matter and when they matter. So the first stage of this book is to identify historic examples which can be used to tease out what really counts. This will make it possible to establish how the evidence of the past can illuminate current predictions of what the economic consequences of climatic change will be. Armed with this insight it may then be possible to decide whether forecasting techniques are able to address the essential features of the economic impact of possible changes and whether the resulting predictions have any utility in terms of taking action.

1.1 Weather or climate?

So far, in the text and in the title of the book the terms 'weather' and 'climate change' appear to have been used interchangeably. This blurring of well-defined concepts is enough to raise the hackles of any self-respecting meteorologist. The reason is simple: we are concerned about a continuum of events which inevitably makes drawing a sharp line difficult. So, we can accept the definition that weather is what is happening to the atmosphere at any given time, while climate is what we would normally expect to experience at any given time of the year on the basis of statistics built up over many years. But when it comes to discussing the impact of extreme events this clear distinction is less easy to maintain.

When the Great Storm of October 1987 wrought havoc over southern England or Hurricane Gilbert tore across the Caribbean in 1988, these were major weather events. But when England was hit by another intense depression in January 1990 or Gilbert was followed by the huge losses resulting from hurricanes Hugo in 1989 and Andrew in 1992, people started to talk in terms of climatic change. The trouble with this translation from weather events to climatic change is, however, that we need to be careful with the statistics. It is all too easy to cite the close succession of a few extremes in some carefully selected period as being significant. If, however, we look at the longer term and take a more rigorous line with the definitions of events, and the timescale over which they are considered, we can produce very different conclusions. For instance, much has been written about the surge in damaging hurricanes in the Caribbean and tropical North Atlantic and its link with global warming. But many commentators paid too little attention to the statistics. Far from showing a dramatic increase, not only has the incidence of tropical storms and hurricanes shown a marked decrease since the 1960s but also the peak winds which cause so much of the damage have declined (Fig. 1.1).[3] Although there is an unresolved issue as to how much of this decline was due to changes in how observations were made, there is no question of an increase in recent decades. Similarly, the incidence of gales over the British Isles has not shown any obvious trends in the last century or so (Fig. 1.2).[4]

The challenge of handling statistics is compounded when the weather events being considered cover whole seasons. Droughts are usually of greatest interest in respect of the growing seasons in mid-latitudes, and in terms of the behaviour of the rainy season in more arid parts of the world. These figures, although the average of several months, can fluctuate wildly from year to year. What constitutes a few exceptional seasons as opposed to a sustained shift in climate may take many years to establish. In recent decades, the drought in the Sahel has to be regarded as the clearest example of climatic change (Fig. 1.3).[5] But many other widely quoted cases of major shifts have to be looked at more carefully. For the

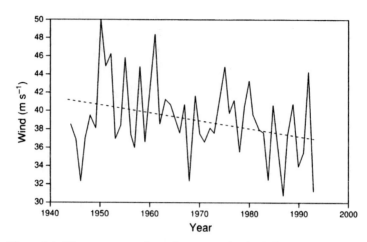

Figure 1.1. The mean annual maximum sustained wind speed attained in hurricanes in the North Atlantic between 1944 and 1993, showing a marked decline since the 1950s. (From IPCC, 1995, Figure 3.19.)

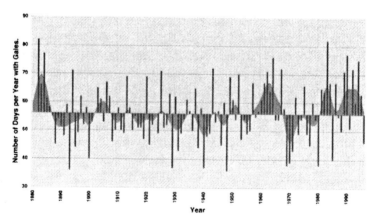

Figure 1.2. The number of days with winter gales in the vicinity of the British Isles over the last 100 years, showing no significant increase in recent decades, together with smoothed data showing longer term fluctuations. (Data supplied by the Climatic Research Unit, University of East Anglia, UK.)

most part they will turn out to be isolated events which may, or may not be, part of a much longer term climatic trend.

The connection between the measurable impact of extreme events and its relevance to calculating the consequences of longer term shifts in the

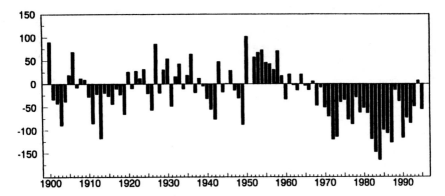

Figure 1.3. Standardised precipitation anomaly in the Sahel Region, showing the sharp reduction in rainfall from the late 1950s onwards. (Reproduced by permission of the UK Meteorological Office.)

climate is an underlying theme of this book. It will continually crop up not only as a statistical issue but in terms of whether the vulnerability of economic and social structures to fluctuations in weather and climate alter with time and what this means for forecasting the impact of future changes. In the meantime, the standard distinction between weather and climate will be maintained wherever practicable. At times, however, there will be a certain fuzziness which reflects the complex interactions between weather, climate and economic activity.

1.2 Fine tuning

The complexity of the interaction between the climate in any part of the world and economic activity can be seen in how societies function. A fall of snow which brings London or Washington to a grinding halt is seen to be of little consequence to Chicago or Stockholm, and positively trivial in Winnipeg or Novosibirsk. Similarly, summer heatwaves which prostrate many temperate countries, are regarded as normal in hotter parts of the world.

The extent to which all aspects of our lives are carefully adapted to the climate becomes even more obvious when we look at different sectors of the economy. Be it agriculture, energy supplies, housing, the retail sector, or the transport system, in their design and operation they all reflect the local climate both in what is normal and in the extremes that can be expected. So, for example, every aspect of agriculture is geared to the local climate and its normal fluctuations. It is only when something truly out of the ordinary strikes that the system breaks down. Even then the response is not passive. Society seeks ways to adapt and so reduce the impact of any repetition of such extreme events.

This variable response to increasingly anomalous weather touches on another essential feature of the economic impact of climatic change. This is the non-linear response of many systems. With modest fluctuations about the norm, systems adapt well and take the ups and downs in their stride. But beyond a certain level they become much more sensitive and can buckle under the strain. What produces the breakdown varies from situation to situation, so it is important to look at a wide variety of past weather-induced social and economic dislocations to get a better idea of how different factors contribute to collapse. This means looking behind these examples to establish what matters most and then exploring whether the messages of the past can be applied to current concerns about climatic change.

1.3 Remembering the past

Because history is rich in fascinating examples of how weather and climate change have played a central part in moulding events, we need to be careful when choosing examples. A selective interpretation of the evidence can lead to contradictory conclusions. What is needed is a sense of proportion in deciding, in the aftermath of some extreme weather event or sustained climatic abnormality, to what extent any outcome could be attributed to the meteorological conditions. So we have to consider what other factors come into play at the time to produce a given outcome, and decide whether in the longer term this outcome was likely to happen in any case.

If we take the example of the failure of the Norse colony in Greenland, this discipline is still likely to lead us to conclude that climatic change played a major part. But, in the case of the decline in the Mayan civilisation in the ninth century we cannot assume that other factors, including pestilence, population growth and war, did not play a more important part in the social collapse – although declining rainfall may have finally tipped the balance. Other frequently cited examples – such as the disappearance of the thriving Harappan culture in the Indus valley around 1500 BC, the dark ages that descended on the eastern Mediterranean at the end of the thirteenth century BC, the Byzantine Empire in the seventh century ad, and departure of the Anasazi from their well-developed communities in Arizona at the end of the thirteenth century – beg similar questions.

These problems are compounded by the fact that in many cases all we have to work on is the evidence of decline. So we cannot draw on economic data, but have only fragmentary historical records to tell the story of what happened. Given the wide ranges of possible causes for the consequences, it is essential to have a framework for assessing the contribution meteorological events may have made. In this respect, the historical approach of defining whether such fluctuations in the weather constituted nothing more than a fateful event or whether they amounted to a crisis or, worse still, a catastrophe, is a useful way to categorise the impact. This qualitative graduation can provide a useful way to discriminate between various examples of the historical and economic impact of climatic change.

A fateful event is where, even though the weather played an important part in the outcome of major developments, it was purely random or coincidental. A good example is the 'divine wind' (*kamikaze*) that destroyed Kublai Khan's Mongol invasion fleet in the Sea of Japan in 1281. There is no doubt that the weather was the dominant factor in the destruction of the fleet. But this was a product of chance, not a sustained shift in the weather in thirteenth-century Japan. If Kublai Khan's astrologers had chosen a different date for sailing, the outcome might have been very different, but this is mere speculation.

By comparison, the contribution of weather fluctuations or climatic change to historical crises is much more difficult to unravel. A *crisis* develops over a sufficiently long period to lead to either a breakdown in social structures and order or a radical change in direction. For this to result from meteorological events would probably require several years, if not decades, of changes in temperature or rainfall to produce, say, a series of poor harvests or crop failures. Even then it is likely that although these meteorological developments might erode the foundations of a society, they would, at most, be only part of the story.

It is not just the complex way in which a crisis builds up which matters, but also how its impact is recorded for posterity. The progress of civilisations are usually marked by the creation of artefacts, monuments and records of successes. During periods of decline these activities are either drastically curtailed or cease. This dramatic response may exaggerate the scale of disruption. So, if, say, a prolonged drought produced sustained food shortages, it would become increasingly difficult to maintain the extravagant and elaborate trappings of an advanced culture. It would also make it more vulnerable to disease, external attack or internal revolt. Where the society suffered internal collapse, it is probable that the surviving population reverted to a subsistence economy, which meant the lot of peasant farmers changed much less. But the surviving records paint a picture of disproportionate change. If this is the case, then it is possible that relatively small but sustained changes in the climate could undermine a social structure rendered brittle by extravagance in times of plenty. Such changes would, however, be difficult to detect and would inevitably combine with other internal and external factors to blur the picture. So any attempts to identify evidence of climate-induced change in the ups and downs of past civilisations must reflect this non-linear response.

By comparison, *catastrophes*, where climatic factors dominate social change, are relatively easy to handle. The only problem is they are exceedingly few and far between. Indeed the collapse of the Norse colony may be the only example of where the deterioration of the climate is the principal cause of collapse. Even here there are plenty of people who are prepared to argue that other factors were equally important. For the rest, it is probably best to work on the basis of either fateful events or extreme weather which lasted long enough to contribute a crisis.

While this classification is designed principally to enable us to consider the riddles of the shadowy ups and downs of distant history, it can be applied to more recent events. So, when we turn to better documented examples of European history, this unravelling of the meteorology from the demographic, economic and social developments becomes even more complicated, given the subtle changes that were taking place. Nevertheless, the part played by the deterioration of the climate in the apocalyptic events of the fourteenth century, or the Tudor inflation, provide an excellent starting point for exploring these issues. They also set the scene for equally contentious questions surrounding, say, the agricultural recession in the late nineteenth century in Britain, the 'dust-bowl' years of the 1930s in the USA or what 'General Winter' played in the defeats of Napoleon and Hitler in Russia. In Steinbeck's description of the drought in Oklahoma in *The Grapes of Wrath* or Sergeant Burgogne's escapades during the retreat from Moscow in 1812, the weather dominates, but was it the sole cause and was it that exceptional?

Only by looking at what the real consequences of the weather were in an adequate range of events can we start to build up a better picture of how meteorological fluctuations combine with other aspects of our society to have profound implications. Once we have formed a balanced view of these more distant happenings, we can then look at recent events in a slightly different light. How various factors combine to dislocate economic structures then becomes a matter of not only the severity of the weather anomaly, but the underlying vulnerability of social structures to both sudden changes or sustained abnormal conditions.

The timescale of events is an essential part of the analysis. We all understand how snowstorms, hurricanes or flash floods can wreak havoc in a matter of hours. Their impact is immediate and often devastating to property and in loss of life. Similarly, the build-up of a prolonged cold spell or a hot dry summer can be appreciated, although the real damage may be less obvious to see. So, while the impact of the drought in Britain in the summer of 1976 had obvious agricultural and water supply implications, the biggest economic impact took place underground. The drying out of clay soils and the damage to the foundations of houses cost the insurance industry over £100 million ($160 million)[6] at the time. Even more insidious and so more difficult to quantify was the resulting increased premiums for insuring domestic properties and parallel stiffening of building regulations to protect new buildings from future drought-induced settlement.

The quoted costs of extreme events depends not only on the extent and form of the damage, but also on who pays. In the case of flood damage in the USA, often properties are not covered by private insurance schemes. Many people rely on Federal cover, but others have no cover and have to fall back on disaster relief if the axe falls. This means widely different figures are quoted depending on who is doing the counting. In the case of the massive Mississippi floods of June 1993, the global figure estimated the damage at $15 billion. But the exposure of the insurance companies was much smaller, being just under $1 billion. Given that agricultural losses were around $5 billion, the remainder was picked up by either the Federal Government or absorbed by individuals. Wherever possible, global figures will be used in this book, and if a breakdown exists of how these costs have been compiled, this will be given.

Assessing the impact of longer term changes requires even more detective work. While the trend may show up in more frequent droughts, floods, hurricanes or storms, the economic consequences are blurred by not only the drawn-out nature of the changes, but also the process of adaptation. This automatic adjustment as part of the continual process of social development and economic renewal also disguises the impact of climatic change. It also provides a reminder not to overstate the consequences of these changes. Since all societies have evolved to adapt to change as an essential part of their development, most aspects of past climatic change will inevitably have been absorbed in this evolutionary process. So we must be careful not to assume a causal connection between climatic developments and other social and economic changes where none may exist. This ability of markets to adjust to substantial changes in supply and demand is central to the arguments of those economists who say that the impact of global warming associated with the build-up of greenhouse gases in the atmosphere, along with other aspects of exponential growth in the consumption of natural resources, has been exaggerated.[7] They maintain that the dynamic response of markets will accommodate the predicted changes. So the question of how fast systems can respond to new challenges is central to the prediction of the future impact of any given climatic change. This depends on how quickly people react to warnings of future environmental threats and how much they are willing to pay to avert them.

1.4 What is the 'real' impact?

These opening observations about the problems of attributing economic consequences to climatic change lead to an even more contentious issue. This is the matter of measuring the real economic impact. It emerges in two particular ways.

First, there is the question of the changing price of goods and services and the consequences of inflation. While correcting figures for the changing cost of living can remove the crude effects of inflation, it is only part of the story. Changing public perceptions of risk and resultant behaviour can have a much greater impact. The huge rise in weather-related insurance claims in recent years may be more to do with the risks people take with their property and the compensation they expect to receive than to changes in the incidence of damaging weather events. Unscrambling this complex interplay has to be part of reaching sensible conclusions about what could be the real consequences of climatic change.

Even more contentious is a second factor – making comparisons between the damage in the developed world and that in the developing world. This issue has come into sharp focus following the Climate Convention of 1995 in Berlin, as national governments have tried to reach agreement on cost-effective policies for preventing the predicted warming of the global climate. The proposal that the economic consequences of loss of life should vary by a factor of 15 between the developed and developing world produced understandable outrage among the representatives of the developing world. Equally, the question of having a sliding scale of permissible increases in the emissions of greenhouse gases to enable less-developed countries to make economic progress in catching up the developed world stirred strong emotions. But, it has to be accepted that in the current world economic order the costs of weather-related damage in developed countries in recent years have been far greater than in the developing world. The situation in respect of loss of life has been the reverse. Providing a balanced account of past climatic impact and calculating the predicted consequences of future changes requires sensitive handling of these issues.

This regional response extends to local and national interpretation of climatic change. If weather statistics in a particular part of the world show no appreciable trend in, say, temperature or rainfall, this is what will guide national governments. Whatever worrying overtones global changes may have, policy will be driven by changes in the national incidence of heat-waves, cold spells, or the strength of the summer monsoon. Global warming only really matters politically if it is driving up the price of bread and potatoes in the local markets. For this reason particular attention will be given to long-term statistical series which provide the clearest evidence of how conditions have changed from place to place. I make no apology for the fact that the figures may look remarkably alike: their similarity contains an important climatological and political message.

All these issues must be addressed in trying to reach conclusions about the economic impact of climatic change. But it must be recognised that there is no agreed weighting that can be attributed to climatic factors in taking wider policy measures and making investment decisions. Moreover, changing the weight given to the various factors influencing decisions can produce radically different results. To avoid getting bogged down on value judgements here the argument will concentrate on the proportionate impact of weather-related events. This approach will consider how the scale of the impact of a given event compares with the normal fluctuations in any specific economic indicator. So where the weather leads to a halving of output, when normal variations amount to no more than a few per cent, then it should be possible to draw conclusions about the vulnerability of the system. This will serve to highlight the non-linear response of the economy and also to underline the adaptability of social and economic structures in the normal range of circumstances. At the same time this approach will bring out the importance of the rate of change in making an impact. Much has been said about how the predicted rate of global warming will be faster than anything that has been experienced in recorded history, and that it is the suddenness of this change which constitutes the major threat to society. There is little doubt that our ability to adapt will depend on the pace of change. So, in looking at the past to draw guidance for the future, the response to this indicator is a more important factor than some measure of the absolute cost of specific single events.

1.5 The point of forecasting

If there are so many pitfalls in identifying the economic impact of past climatic change, then it would hardly seem unreasonable to question the point of making any forecasts. Weather forecasting has long been the butt of public ridicule. Economic forecasting has an even less enviable public status. So the combination of these two specialist activities is liable to produce something which is held in still less regard and hence lead to the conclusion that the whole exercise is a waste of time.

The response to this calumny does not lie in trying to justify the performance of forecasts or the progress that has been made of late, but is found in the fundamental nature of decision-making in a democratic society. Because those who make decisions are normally accountable for their decisions to, say, elected representatives or shareholders, there is a requirement to show that they had taken adequate account of the options and how these may be influenced by future developments. Since the success or failure of policy or business decisions inevitably depends on making reasonable judgements about the future, it involves making use of forecasts. Whatever the limitations of such predictions, the defence of both those who issue them and those who use them to make decisions is that they were the best that could be produced at the time. The alternative of disregarding the accepted view of future developments is far less easy to justify.

These decisions occur in all aspects of private sector and public sector business. So, for instance, US energy utilities must take short-term views of natural gas prices as a hurricane threatens off-shore installations, or distribution networks must learn from bitter experience when Hurricane Andrew did so much damage in Louisiana in 1992. Looking further ahead, such companies must take a view of future peak demand for natural gas, heating oil and electricity in extreme cold spells and make investment decisions years in advance to be prepared for the demand generated by growth in the economy. To disregard forecasts of both potential demand and the probability of extreme weather would be regarded as folly, especially if, with the benefit of hindsight, cheese-paring decisions had been found wanting. The same rationale applies to other industries and in other countries.

Once this basic reliance on forecasting is accepted as an essential part of making business decisions and setting economic policy, then it follows that the best approach to forecasting is to be realistic. This involves understanding the principles it is built upon, recognising the limitations it operates under, and finding ways of refining its output. In the case of weather and climate predictions and their potential application to improving economic policy formulation, the challenges are massive. The potential benefits are, however, just as substantial. Central to making sensible decisions on the basis of forecasts is drawing effectively on the past. At one level, what is needed is a proper understanding of the probabilities of an extreme event using climatological statistics. It is more illuminating to ensure that our plans address the weaknesses in economic and social structures exposed by past experience to enable us to be reasonably confident that we are addressing the right issues.

The lessons of the past help in addressing the limitations of forecasts in various ways. By their nature long-term predictions can only give broad indications of future trends. So climate forecasts predict that temperatures will rise by, say, two to four degrees Celsius in the next 50 to 70 years if the concentration of greenhouse gases rises to certain levels over this period. Similarly, economic forecasts of future rates of growth are built on broad assumptions about the availability of various basic resources. In practice, because these climatic forecasts include more rapid alterations in the rate of change than in the past, they imply a future containing more surprises.

The impact of the pace of change on extreme events is crucial when considering forecasts of global warming of a few tenths of a degree Celsius a decade. It is not difficult to dismiss the broad increase in temperature for two reasons. First, technological optimists can argue that we take a gradual warming in our stride. It can be denied on the grounds of being no more than the equivalent of the difference that has been created over the last century in major urban areas or exists naturally in adjacent geographical regions. Secondly, there is the issue of winners and losers. To some parts of the world additional warmth would reduce heating bills in winter and improve agricultural yields. The hot, dry summer of 1995 in Britain, seen by many as a foretaste of life in the greenhouse, was a boon to farmers as it followed a mild wet winter, so there was adequate soil moisture to enable cereals to mature early, reaching their full potential and allowing them to be harvested in ideal conditions. Accurately assessing who gains and who loses around the world is an essential feature of useful economic forecasts. These figures are needed to inform the debate on the balance of advantages and disadvantages of change, and also for international negotiations on how individual countries should contribute to achieving targets for reducing the climatic impact of human activities.

These geographical ups and downs must be combined with the inevitable temporal fluctuations. As will become clear, any forecast trend in climatic change is small compared with the changes from year to year. So what will matter most is whether certain extremes will become more frequent as part of any climatic change, as it is these that will impose the greatest burden on the economy. This emphasis on extremes becomes even more important if the assumption of a progressive change in the climate proves incorrect. It is fully within the bounds of possibility that the non-linear nature of the climate could lead to a sudden shift in global weather patterns. What is more, while the most probable response to human activities is to produce a warmer world, there is a smaller but finite possibility it might flip into a colder regime for some parts of the world. If a more dramatic response is on the cards, then some form of extreme will become the order of the day; but which one will be in the lap of the gods. So, as the first stage in exploring the economic impact of climatic change, we must look closely at the extremes of the past and consider how they have conspired with other events to have a major influence on human history.

In conducting this exploration of past events and their economic and political implications, I will do my best to provide a truly international picture. But the fact that I have lived and worked for most of my life in England is bound to influence how I interpret the issues. Furthermore, the Old World, and Europe in particular, has the most extensive quantitative historical records which enable us to tease out economic observations, and this means that the potential for geographical bias is considerable. Nevertheless, the objective is to identify generic messages which apply more widely. So, while the economic analysis will be underpinned by UK experience, its goal is to find global messages.

1.6 Notes

1 The quotations at the head of each chapter come Lewis Carroll's *Alice's Adventures in Wonderland* (first published in 1865) and *Through the Looking-Glass* (first published in 1872).

2 Maunder (1970; 1987).

3 Landsea (1993).

4 Hulme & Jones (1991).

5 IPCC (1990).

6 Where figures for losses are quoted in pounds sterling (£) the conversion into US dollars ($) is given at the rate of £1.0 = $1.60, which is the rate prevailing in early 1997, although at the time the losses occurred a different value may have applied. This conversion is only used for figures since the early 1970s, when flexible exchange rates became the norm. Prior to 1970, no conversion is given.

7 This point of view is expressed in its most unexpurgated form in Beckerman (1995). This pure economic approach – 'red in tooth and claw' – infuriates many ecologists, who argue that an economic price cannot be put on the ecological, moral and social assets which are put at risk by global warming. This is the dialogue of the deaf. It will not be resolved by intellectual debate but by which point of view is endorsed by voters in choosing their elected representatives. Here we can only note the objections to the economic approach, while recognising that broader arguments exist. But it does have to be said that, when it comes to the crunch, the majority of voters still seem to be lured by the economics of self-interest, and this political reality will influence the analysis presented here.

2

The historical evidence

'Where shall I begin, please your Majesty?' he asked.
'Begin at the beginning', the King said, very gravely, 'and
go on till you come to the end: then stop.'
Alice's Adventures in Wonderland, Chapter 12

For us the beginning has to be the role of climatic change in the rise and
fall of ancient civilisations. This is the subject of heated debate among
archaeologists, climatologists and historians.[1] The kernel of this debate
has been the lack of reliable evidence of what precisely happened to the
climate. While the broad sweep of changing conditions can be inferred
from available data and conclusions can be drawn about rises and falls in,
say, rainfall, translating this into explanations of the success or failure of
societies is more problematic. So the fragmentary clues of what led to
collapse have often been open to a variety of interpretations. This has led
to a polarisation of views which was not always helpful. What lay at the
heart of the matter was failure to appreciate how limited the knowledge
of the real nature of past changes in the climate was. In recent years, an
increasing amount of evidence from proxy climatic records has been collected from a wide variety of sources (e.g. tree rings, lacustrine deposits
and pollen records). These new data are throwing a different light on the
causes of waxing and waning of ancient civilisations.

2.1 Lost civilisations and dark ages

Lacunae in orderly progress of history, punctuated as it is with dramatic
collapses in civilisations and blank periods where no records exists, exert
a peculiar fascination over historians. The picture of ancient civilisations

is built up principally from permanent records that have survived, together with lasting artefacts and buildings studied by archaeologists. Where these are missing the combination of mystery and frustration provides fertile ground for theorising on the causes of decline and fall. It is widely recognised that the interruptions in the permanent records and the fall of dynasties and empires may exaggerate the changes for the majority of the populace – the subsistence economy of the peasants continued as they eked out a basic existence come what may. Nevertheless, the sudden collapse of societies begs explanation, and there is no shortage of theories to fill the gaps. The principal shortcoming with many of these explanations is the desire to find a single solution to the riddle. Here, in discussing the part played by either extreme weather or sustained climatic change the emphasis will be on how these events combined with other factors by chance to bring about economic and social collapse.

The first example of this process is the disappearance of the civilisation that thrived in the Indus valley between 2500 and 1500 BC (Fig. 2.1). The scale of the cities in the Indus valley (Harappa and Mohenjo-daro) and the suddenness of their collapse have led to much speculation as to what could have caused a disaster. The absence of evidence of warfare or rebellion, and the total abandonment of the sites is an open invitation to seek a climatic explanation. What evidence there is does suggest that between 1800 and 500 BC rainfall across the Middle East and into northern India declined appreciably. So this may be part, if not all, of the explanation of the decline of the Harrapan civilisation.[2]

The same lengthy period of desiccation can also be invoked to explain the even greater puzzle of the dark age that descended upon the eastern Mediterranean at the end of the thirteenth and beginning of the twelfth century BC. This period saw the collapse of the Hittite empire, Mycenae and Ugarit and the enfeeblement of Egypt. Usually attributed to the influx of warlike 'Sea Peoples',[3] there is no clear explanation of what drove these movements and why they had such a disruptive effect. Indeed the pictorial record suggests that Ramesses' 'famous victory' over these marauders was not against a well-ordered military force but the result of a mass movement of people including women and children fleeing from some greater disaster (Fig. 2.2). While a climatic explanation is not accepted by many historians, the coincidence of an agricultural crisis, due to a prolonged decline in rainfall, could have magnified the disruption caused by other factors. Suffice it to say many of the Bronze Age people of the region relapsed back into a subsistence economy for two to three centuries before

Figure 2.1. A set of seals from the Harrapan (Indus) civilisation depicting various animals. (From Sheratt, 1980.)

the Iron Age brought the flowering of Greece. The tenuous thread through the dark age between these two periods is recorded in the legends of the Trojan War.

Equally tantalising and more surprising is the dark age of the seventh century AD in the Byzantine Empire. Although the events of this period are better documented, the scale and nature of the decline are often overlooked. Here the explanation has to be multifaceted. Although there were signs of social stress in the Byzantine Empire in the early sixth century,

Figure 2.2. A detail of the land battle between the Egyptians and the 'Sea Peoples' *circa* 1186 BC from the relief on the Great Temple of Ramesses III at Medinet Habu depicting the enemy force as including wicker-sided ox carts with women and children in the melee. (Reproduced by permission of Dr Nancy Sandars.)

possibly as a consequence of population pressures, the first hammer-blow came with the arrival of bubonic plague from Ethiopia in 542. It raged in Constantinople during the spring of 542 and may have killed as much as a third to half of the population. It then returned with terrifying regularity, every 15 years or so, to most of the major cities. By the end of the sixth century the scale of depopulation was horrendous, with many cities which had survived since Antiquity ceasing to exist. In effect much of the Mediterranean world slipped back into a form of rural convalescence, with life continuing more easily in the countryside where contagious diseases exerted a less deadly sway.

The historical implications of these changes bear examination. The Byzantine Empire continued to exist, and indeed expanded briefly in the early seventh century under Heraclius, who won a crushing victory over the Persians at Nineveh in 628. But this encounter may have exhausted both sides, and it has been argued[4] that in many places Islam effectively expanded into a vacuum as Arabia had been spared the effects of plague. While Constantinople survived, in spite of being blockaded by the Arabs in 674 to 678 and put under siege in 717 and 718, by the mid-eighth century the population had sunk to between 25 000 and 50 000 compared with a figure of some ten times this at the beginning of the sixth century. From the mid-eighth century things began to improve. Even so, the Arab geographer Ibn Khordâdhbeh stated in 841 that there were only five cities in Asia Minor – Ephesus, Nicaea, Amorium, Ancyra, and Samala (?) – in addition to a considerable number of fortresses.[5] During the dark age trade and economic activity virtually died out. Archaeological studies at Sardis, the capital of Lydia, showed a dramatic decline in the use of bronze coinage (small change which is a useful measure of economic activity). The years 491 to 616 were represented by 1011 bronze coins, the rest of the seventh century by about 90 and the eighth and ninth centuries by only nine. Similar results have been obtained in nearly all Byzantine cities.[6]

It has been argued that climatic change played a part in this extraordinary decline.[7] Indeed a more cataclysmic event may have triggered this process. There is widespread evidence of what is usually assumed to have been a massive volcanic eruption in 536, possibly of Rabaul in New Guinea. Chroniclers from Rome to China record that the Sun dimmed dramatically for up to 18 months and there were widespread crop failures.[8] The creation of a global dust veil in the upper atmosphere would have absorbed sunlight at high levels with a corresponding cooling at the Earth's surface. There is evidence of a significant deterioration in the climate for several years in tree ring data from north-west Europe,[8] but nothing clearly relating to the eastern Mediterranean. So without better information about whether the weather conditions altered for long enough and sufficiently to contribute to the Byzantine dark age, it is not possible to draw any definite conclusions.

A recent example of this process is a study that comes from Central America.[9] Analysis made of a sediment core taken from Lake Chichancanab, in what is now Mexico, provided a continuous record of how local conditions varied over the last 8000 years. By measuring the oxygen isotope ratio in the calcium carbonate of the buried shells of the shellfish that lived in the lake, it is possible to establish the dryness of the climate over the years.[10] From these measurements the researchers concluded that the driest period in the last 8000 years was around 750 to 900 AD.

This result may hold the key to one of the great mysteries of prehistoric Mesoamerica: the sudden collapse of the classical Mayan culture. Having emerged around 1500 BC this civilisation thrived from around 250 AD. It is renowned for its monumental constructions and reached a pinnacle in the eighth century. By this time the population density in the Mayan lowlands (which extend over modern-day Guatemala, Belize, Honduras and Mexico) was far higher than current levels. It sustained a sophisticated society which built magnificent buildings and other edifices. But early in the ninth century the civilisation entered a cataclysmic period, although all the other social and demographic pressures had already been in operation for some time. The monumental construction and detailed records (Fig. 2.3) came to an abrupt end at one centre after another.

Archaeologists have long argued about the causes of this collapse. But none of the proposed explanations of overpopulation, epidemics or invading armies seemed to provide an adequate explanation for the suddenness of the collapse. Perhaps the most plausible is proposed by Sir Eric Thompson[11] that peasants revolted against the overbearing demands of the massive theocracy. This would have suddenly reduced society to a subsistence economy with no resources devoted to producing permanent artefacts. The new data show how the improved climatological information can provide new insights into the possible causes of such changes. To a civilisation facing a number of internal pressures of the type proposed by Sir Eric Thompson, a period of drought would have greatly increased the susceptibility of many Mayan cities to revolt. This would also explain why the cessation of records at many cities varied over about a century. Spread over two centuries, only those centres with riverside locations could have survived sustained drought, which is consistent with the pattern of decline. So the new data provide substantial support for the hypothesis that climatic change was the principal cause of the collapse of this enigmatic civilisation.

By comparison with the examples covered so far, the disappearance of the Norse colony in Greenland seems an open and shut case. But, even here, the data need to be treated with care. The historical records of contacts with the colony, together with data from the ice core drilled in the Greenland ice sheet, provide a pretty clear picture of climatic deterioration.[12] What is still the subject of discussion is whether this was sufficient to destroy the colony or whether other factors need to be considered to explain its inability to adapt to changing circumstances. In particular, the failure of the community to take on Inuit technology to accommodate the

Figure 2.3. An example of a Maya commemorative sculpture. It stands well over 7 metres high, and according to its text was erected in 761 AD. (From Sheratt, 1980.)

worsening climate may be an important example of how social rigidities amplify the effects of climatic change.[13]

2.2 Born to woe: The calamitous fourteenth century

Having considered the broad sweep of early history we can turn to the more thoroughly documented history of western Europe. While this provides the reassurance of more quantitative data, we must not lose sight of how many factors are at work and how they combine to steer events. And, where better to start than the apocalyptic events of the fourteenth century. This period when famine, war and plague ravaged Europe seems to cry out for climatic change[14] to be part of the explanation. Moreover, there is evidence that these upheavals were part of a global decline. In China the estimated population fell by about 40 per cent from a peak of around 100 million in the mid thirteenth century in the following hundred years or so.

There is plenty of evidence to suggest in Europe that not only were the eleventh and twelfth centuries marked by a period of benign climate – often termed the 'medieval climatic optimum' (see Section 5.3) – but also there was a marked deterioration of the weather during the thirteenth century. Increased storminess in the North Atlantic had largely cut communications with the Norse colony in Greenland. This greater storminess also brought more frequent disastrous inundations of low-lying coastal areas around the North Sea.[15] In North America the Anasazi abandoned their well-established communities in the desert south-west USA at the end of the thirteenth century, while in Wisconsin the northern limit of maize cultivation receded southwards by up to 320 km.[16] So by 1300, the climate had taken a turn for the worse: what is less clear is whether this was a global cooling and, if so, what part it played in the economic and social calamities that hit so many parts of the world. The data from ice cores in Greenland show an upsurge in volcanism around the middle of the thirteenth century. A major unidentified eruption in 1259 must have had a significant, if only temporary, cooling effect.[17] Thereafter, volcanic activity seems to have remained at a much higher level than in the previous two to three hundred years, especially between 1285 and 1295, and in the 1340s, which could have contributed to a general cooling of the global climate.

Whatever the causes, the deterioration in the weather in Europe seems

to have been well established by the year 1300 and the new century got
off to a bad start. Two exceptionally severe winters gripped northern
Europe in 1303 and 1306, then between 1314 and 1317 there was a run
of extraordinarily wet and cool summers. The disastrous harvest failures
of these years come ringing down through history as the greatest weather-
related disaster ever to hit Europe. From Scotland to northern Italy, from
the Pyrenees to Russia, an unequalled number of reports exist of the awful
consequences of the dreadful weather.[18]

In England, the harvest of 1314 had near normal yields but wet weather
made the harvest difficult. But it was in 1315 that the rain started in
earnest at Pentecost (May 11) and continued almost unceasingly throughout the summer and autumn. The harvest was disastrous. Even in the
fertile land of the Bishopric of Winchester the yield was less than two and
a half grains for one sown (Fig. 2.4). In London by the early summer of
1316 the price of wheat had risen by as much as eightfold from that of
late 1313, reaching levels that were not equalled again until the late fifteenth century. Legislative attempts to hold down prices failed, although
concerted efforts were made to control the price of ale. This was particularly costly for brewers who had bought grain at high prices and could
not recoup their investments.

The same story was repeated in France and had a wider impact. A
campaign by the King, Louis X, to bring Count Robert of Flanders to heel
was frustrated by the sodden weather. The invading army was brought to
a complete halt in the waterlogged Flemish countryside. Horses sank up
to their saddle girths and wagons became bogged down so that seven
horses could not move a single wagon. Within a week Louis gave up and
withdrew in disarray.

The constant rain, low temperatures and dark skies not only spoilt the
grain, but also prevented the usual production of salt by evaporation in
western France and produced a tiny, sour, late wine harvest. So with all
essentials in short supply, famine and pestilence stalked the Continent,
from the autumn of 1315 onwards. At Louvain grain prices rose more
than threefold between November 1315 and May 1316. The filthy weather
also caused large numbers of sheep and cattle to succumb to various diseases.

Starving peasants were reduced to eating dogs and frogs. There were
widespread reports of cannibalism, with mothers eating children, graves
being robbed and the bodies of criminals cut down from gibbets to be
eaten. Such grotesque reports are commonplace in many reports of other

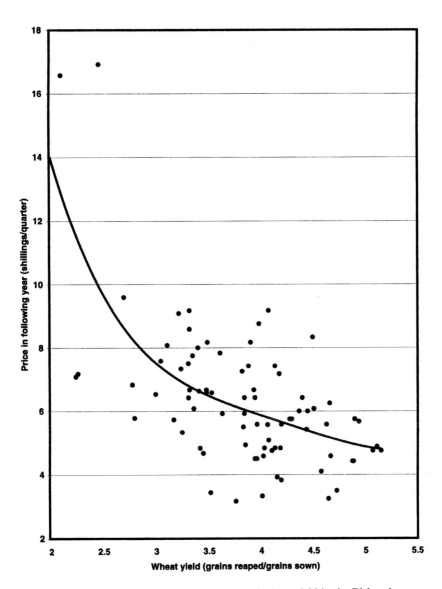

Figure 2.4. The correlation between prices and wheat yield in the Bishopric of Winchester during the period 1221 to 1350. (Data taken from Titow, 1960.)

famines in western Europe before and after 1315. There is, however, little
doubt the number of reports singles out this famine as exceeding all others
in its severity and scale across Europe. Mouldy grain resulted in outbreaks
of ergotism and the dangerous skin disease erysipelas (St Anthony's fire).
The weakened populace died of a wide range of ill-defined fevers and
murrains or simply starvation. The death toll is hard to gauge. In isolated
communities it may have been very high, starting the phenomenon of
deserted villages, two generations before the Black Death was to have a
far greater impact. In the advanced city of Ypres, remarkably detailed
records suggest that over 10 per cent of the population died during the
summer of 1316.

The harvest in England in 1316 was even worse but further south
things were better. Together with imports from southern Italy, which had
escaped the meteorological crisis, the picture gradually improved. Though
the severe shortages still existed in 1317, the worst was over. But in eastern Europe and in Ireland the agony dragged on for another year. Even
allowing for the vulnerability of medieval society to bad weather, the
events of 1315 and 1316 seem to match anything in subsequent years.
They probably represent the extreme example of the weather in northern
Europe being dominated by an unbroken stream of depressions from the
Atlantic.

Thus far the case for climatic change exerting a profound influence on
the fourteenth century seems to be growing. But, before we get carried
away, there are two important factors to be brought into the discussion.
First, by around 1300, there was already widespread evidence of demographic pressures and this may exaggerate the impact of weather. Marginal lands were being occupied for arable farming, and the balance was
shifting between raising livestock and growing cereals, causing problems
of declining yields and making them more vulnerable to adverse weather.
The high mortality reported as the result of inundations by North Sea
storms may be another aspect of these pressures in that people were prepared to take more risks to exploit available land. The decline in population following the famines of the 1310s and the widespread desertion of
marginal lands were the first stage in redressing the balance. So, well
before the Black Death arrived, the fourteenth century was already a Malthusian disaster waiting to happen.

The second factor is that after its awful start, bad weather does not
feature prominently in the reports of the terrible events of the remainder
of the fourteenth century. In fact, it does not seem to be that much out

of the ordinary. We have the benefit of the earliest weather diary prepared by the Reverend Father Merle, mostly at Driby in Lincolnshire, but also on visits to and from Oxford between 1337 and 1343.[19] The description of weather events reads very much like twentieth-century experience. Moreover, while the heat of the summer of 1348 has been invoked to explain the spread of the Black Death in England, in general the weather does not feature prominently in the litany of woe that punctuated the remainder of the century. So its contribution to the course of the initial epidemic of the plague or subsequent waves of pestilence and war that lapped over the subsequent hundred years or so seems at most marginal.

Where the weather may have played a more important if shadowy role is in the genesis of the plague. It is widely assumed that this lay in the horrendous floods that devastated China in 1332. Reported to have killed several million people, they caused huge disruption of large parts of the country and substantial movements of wildlife, including rats, in which bubonic plague was endemic. It is now recognised that the mixing of different populations of rats following major natural disasters is a crucial factor in triggering new outbreaks of the plague. So it is reasonable to conclude that the Black Death pandemic was triggered by the conse-quences of massive floods which struck China in 1332. Once this virulent new strain of the disease had emerged, its subsequent spread was con-trolled by events which were largely unrelated to weather and climatic change.

This first glimpse into the murky subject of using more complete his-torical records to provide a better insight into the economic consequences of climatic change established many of the ground rules of the subject. First, it indicates how certain records of agricultural activity and the prices of products contain a lot of information about economic activity. This is hardly surprising as in the Middle Ages roughly 80 per cent of working class expenditure was on food and drink, so ups and downs in cereal prices represent the heartbeat of the medieval economy. Since these were closely related to the quality of the harvest (see Fig. 2.4) they are directly related to the weather during the growing season.[20] But, as has already become apparent, the nature of the relationship is not simple and frequently is ambiguous.

In spite of these limitations, price data and other records of agricultural activity are a valuable source of meteorological information prior to instru-mental observations. More importantly, they contain the essence of the economic activity of the societies we wish to find out more about. So the challenge is to make use of the records without falling into the trap of reading too much into the tantalising features that shimmer mirage-like on their surface. This means that, whenever looking at given sets of records which appear to contain a clear message of climatic impact, it is essential to cross-check with as many other sources as possible to see whether they tell the same story. Where they do not tally then it is not permissible to be selective in concentrating too much on the data which support a preferred hypothesis.

As we advance towards the present, the opportunities to conduct these cross-checks will multiply. But, the awful events of the fourteenth century are a good place to start as they bring out the complexities from the outset. At one extreme we have already witnessed the efforts to manipulate or control prices: a political inclination that remains undimmed nearly 700 years later. At the other end, there is the temptation to attach particular weight to certain reported weather extremes. For example, bitterly cold winters are frequently cited as contributing to shortages. As will become clear, in respect of the basic cereal production they are not a crucial factor. Indeed, hard winters seem more likely to have been harbingers of a good harvest than vice versa.

Spring and summer weather is equally hard to categorise. Successful cereal growing in north-western Europe is a fine balance between adequate moisture and reasonable warmth, especially at harvest time. The combination of heat and drought can be as damaging as sustained cool, wet weather. On balance, cool relatively wet summers produce the heaviest yields, providing the harvest can be gathered in efficiently.[21] The recipe for success tends to be plentiful rainfall and average temperatures, until the end of June, then reasonably dry and warm thereafter. Given that wet growing seasons usually feature below average temperatures, these ideal conditions are rarely achieved, but the adaptability of agriculture means that only when certain adverse combinations of weather occur do yields fall dramatically. What constitutes these damaging conditions will be explored as the historic examples of the impact of climatic change are examined. This will not simply be a matter of understanding past events. More important is being able to predict confidently the causes of fluctuations in crop yields. Without this information there is little hope of predicting how food supplies might be affected by changes in temperature and rainfall using global climate models (see Section 6.4).

In parts of the world, which have hotter summers, the situation is less complicated. As we will see in the examples of North America and India,

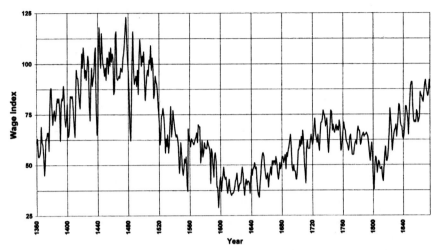

Figure 2.5. The index of the purchasing power of builders' wages in England over six centuries. (Data taken from Phelps-Brown & Hopkins, 1956.)

it is drought which really matters – hot, dry summers are bad news. Conversely, wetter, cooler summers usually produce the best harvests. So global warming looks much more likely to reduce yields in hotter parts of the world.

2.3 The Tudor inflation: Malthus, meteorology or money?

The profound check on population pressure brought about by the Black Death, and sustained by subsequent bouts of the plague, reduced the pressure on agricultural resources for some 150 years. Although the fifteenth century was not plain sailing and there were harvest failures, the recorded incidence of famines was lower than in the late thirteenth and early fourteenth centuries. This does not mean that the climate was quiescent during this period. Indeed Hubert Lamb[22] has identified the 1430s as being a decade that featured an extraordinary number of savage winters in Europe.

The relative abundance of the fifteenth century is best seen in the monumental work of Sir Henry Phelps-Brown and Sheila Hopkins on wages and prices.[23] This shows (Fig. 2.5) that the purchasing power of wages, as represented by those paid to building craftsmen, rose in the second half of the fourteenth century and remained at high levels until the first decades of the sixteenth century. They then fell steadily to reach a nadir in the 1590s and then rose slowly, but did not return to fifteenth century levels until late in the second half of the nineteenth century. There were occasional sharp rises in prices associated with poor harvests which led to a comparable drop in the wage index, notably in 1439 and 1482. The overall picture is, however, of underlying price stability, all of which makes the fivefold rise in prices that started in the 1520s and lasted for around a hundred years so intriguing to economists. Analysis of prices and wages in France has produced a similar picture.

Known as the 'Tudor Inflation', the rise in prices during the sixteenth century has been variously attributed to demographic pressures and to the influx of gold and silver from the Americas which inflated the money supply. But the fact that this development coincided with a marked cooling of the climate which is often referred to as the Little Ice Age means that the fluctuations in the weather need also to be considered. The global significance of this climatic cooling will be reviewed in Chapter 5, but the effects in north-western Europe are well documented. These include an increasing number of reports on the weather as well as an increasing range of economic and demographic series.

The information on the weather is augmented by one particularly valuable series. This is the analysis of wine harvest dates in northern and central France, Switzerland, Alsace and the Rhineland prepared by Emmanuel le Roy Ladurie and Micheline Baulant.[24] This spans the period 1484 and 1879 (Fig. 2.6) and provides an unrivalled insight into the temperatures during the summer half of the year (April to September). While the longer term variations have to be viewed with caution, as fashions on the sweetness of wines changed over the years, and with it the date of harvesting, the fluctuations from year to year provide an accurate measure of the weather as they correlate closely with change in temperature. Moreover, because rainfall is inversely correlated with temperature during the growing season in north-western Europe, so early harvests signal a hot and dry summer, while a late harvest signifies a cool wet season.

An additional value of the wine harvest data is the insight it provides in interpreting the more prolific cereal price records. Because poor harvests, and hence high prices, could be the product of either low temperatures and excessive rainfall or, less frequently, drought and heat, comparison of the different series can sort out the two extremes. The Beveridge European wheat series from 1500 to 1869[25] and the Hoskins series for England from 1480 to 1759[26] are of value in identifying the good and bad

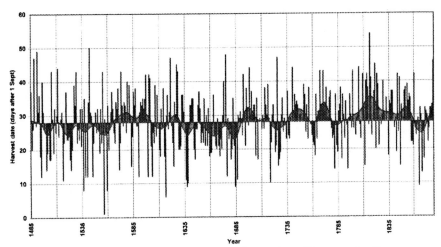

Figure 2.6. The date of wine harvests in northern France and adjacent regions between 1484 and 1879, together with smoothed data showing longer term fluctuations. (Data taken from Le Roy Ladurie & Baulant, 1980.)

years (Fig. 2.7). Furthermore, when combined with records of births, deaths and marriages, such as those in England which stretch back to 1541,[27] it is possible to build up an increasingly comprehensive picture of the interplay between climatic change, agriculture, and society at large in Europe from the beginning of the sixteenth century onwards.

One further set of records deserves particular mention. This is the work of Christian Pfister at the University of Bern. A thorough search of archives has produced a detailed picture of the climate of Switzerland between the early sixteenth century and the early nineteenth century, after which adequate instrumental records are available. The picture has been built from direct weather observations, phenological records and other agricultural information, such as wine harvest dates and yields, plus outstanding events such as snow amount and cover, and the freezing of lakes. The information has been organised into a series of indices for seasonal temperatures and wetness for the period 1525 to 1979.[28]

What emerges from these records is that there was no obvious climatic deterioration until around 1560. Indeed, if anything, up to then there were more frequent hot summers and an unremarkable incidence of cold winters. While there was a period of high prices at the end of the 1520s associated with the cool wet years of 1527 and 1528, the most dramatic figures occur in the late 1550s and were principally a result of the drought

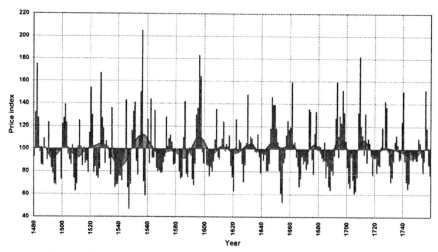

Figure 2.7. The annual wheat price index for England between 1480 and
1759, together with smoothed data showing longer term fluctuations. (Data
taken from Hoskins, 1964; 1968.)

and heat of 1556. The impact was less profound in Europe, but in England
the price of wheat more than doubled – the greatest rise above trend in
the Hoskins series. This followed a poor harvest in 1555 when the weather
was cool and damp, and precipitated a major subsistence crisis in England.
The death rate rose sharply in late 1556, and the combination of high
prices, famine and epidemics of disease pushed it up to well above twice
the trend value in 1558/59: an excess that outstripped any subsequent
crises by a factor of three.

In using mortality figures it is important to appreciate the inevitable
lag between cause and effect in interpreting mortality statistics during and
following periods of famine. Because infant mortality was a major part of
the fluctuations in the death rate in these times, and these deaths are
strongly dependent on problems that occur during pregnancy, there is a
delay between food shortages and peak mortality. High wheat prices
during pregnancy lead to severe infant mortality, but have little impact
on children that survived more than one year. So often the most dramatic
death rates are well after the agricultural disaster and this can easily dis-
guise the role of the weather in events.

Beyond 1560 a different picture emerges. Frequent cold winters and
late wine harvests are seen as clear evidence of a climatic deterioration in
north-western Europe. At the same time, glaciers in the Alps grew appre-

ciably, which is regarded as confirmation that the summers became cooler and wetter. But, in spite of more late wine harvests, neither the wheat price indices nor the wine harvest dates show a consistent picture of sustained climatic deterioration which can be invoked to explain the inflationary pressures of the 1570s and 1580s.

More intriguing, especially in England, is the sustained awfulness of the harvests in the mid-1590s. Between 1591 and 1597 all the wine harvests were late and poor harvests drove wheat prices well above trend for four years in a row. The Phelps-Brown & Hopkins wage index reached its nadir in 1597 when the purchasing power of the earnings of building craftsmen fell to barely a quarter of the best levels experienced in the fifteenth century. The consequence of this dramatic reduction in the standard of living was panic legislation. Parliament passed a Great Act codifying a mass of sectored legislation and local experiments on poverty relief. It also restored many of the restrictions on enclosures of common land and the conversion of arable land to pasture which had been repealed only four years before.

At the beginning of the seventeenth century the Tudor Inflation began to slow and effectively came to a halt by around 1630. Thereafter prices rose much more modestly and real incomes rose slowly, although this progress was punctuated by repeated reversals. As for the reasons for the inflation, two quotes from Phelps-Brown and Hopkins' work set the scene well. First, they observed 'For a century and more, it seems, prices will obey one all-powerful law; it changes, and a new law prevails; a war that would have cast the trend up to new heights in one dispensation is powerless to deflect it in another.' Then of the 1590s they asked: 'Do we see here a Malthusian crisis, the effect of a rapid growth in population impinging on an insufficiently expansive economy . . . ?'[23]

These questions point to a possible explanation. As in the early part of the fourteenth century, the combination is of steadily rising demographic pressures being projected into a major crisis by sustained adverse weather. But this is not enough to explain the cause of the sustained inflation throughout the sixteenth century which is absent in the centuries before or after. The monetarist explanation of the impact of the huge influx of South American gold and silver into the European economy is the essential additional ingredient. It ended the famine of precious metals that had strangled the European economy in the Middle Ages. This both stimulated economic production and also financed imports from the Baltic area, Russia and the Orient. It also inflated the money supply, and concentrated wealth in the hands of the rich. Whatever the economic balance between demographic pressures and monetary expansion, one thing is clear: the role of climatic change cannot be invoked to resolve the debate about the differences in the changes in the fourteenth and sixteenth centuries. Conversely, there is no doubt the fluctuations in the weather from year to year dominated the short-term ups and downs in prices, and with it the whole quality of life for the great mass of the population.

2.4 Subsistence crises of the seventeenth to nineteenth centuries

After the dismal end of the sixteenth century, the early decades of the seventeenth century were mercifully spared of really bad harvests. In the Beveridge and Hoskins series between 1598 and 1629 only 1608 stands out as a notably bad harvest. This year seems to be a case of an exceptionally cold winter and a poor summer combining to produce low yields. Other sources suggest that in the north and west of Britain 1623 was also a year of dearth. In the English mortality statistics a marked peak occurs in 1625, but this has more to do with it being a plague year in London than to a poor harvest. Furthermore the notably late wine harvests of 1621, 1627 and 1628 are not mirrored in the wheat price indices. High prices in 1630 appear to have been the product of a warm dry summer to judge from the wine harvest. Conversely, the three outstandingly hot summers of 1636 to 1638 had little impact on prices. In England the sharp rise in mortality rates in the summer of 1638 may have been related to the spread of dysentery in the hot weather.

In the second half of the 1640s a series of poor harvests pushed prices up, but the wine harvest dates do not show any outstandingly poor summers. A set of abundant harvests followed in the early 1650s, while the latter part of the decade had a poor run, with prices rising particularly sharply in 1661. The climatic causes are, however, unclear.

At this point we can introduce a new source of information. This is the Central England Temperature series produced by the late Professor Gordon Manley.[29] This record of monthly temperatures for rural sites in central England is the longest homogenous record in the world. It runs from 1659 to 1973 and is now regularly updated by the UK Meteorological Office.[30] While its early figures are built up principally from weather diaries, by a combination of skilled detective work and scholarship Manley has woven together a combination of instrumental and other observations to create an internationally renowned series. How this record underpins the analysis of climatic change since the mid-seventeenth century will be discussed in Chapter 5. For the moment, its importance is that it provides a temperature series which can be used to check the meteorological aspects of subsistence crises. Furthermore, it is joined by other series from the early eighteenth century onwards.

The remainder of the 1660s were marked by mainly good harvests, and while the 1670s had the occasional bad harvest, notably 1673 and 1678, the situation was manageable. Similarly, the 1680s were marked by a series of good harvests with 1684 and 1686 being notably hot and dry. The striking feature of these decades is that whereas the growing seasons appear to have been good, for the most part, the winters were notable for their severity. Both in England and Holland evidence clearly points to markedly colder winters, with years like 1676 and 1684 standing out.[31] At the same time, severe winters were interspersed by very mild years. For example, 1686 appears to have been as mild as anything in recent years.

This relatively benign story is brought to an abrupt halt by the 1690s. The decade featured an unparalleled combination of cold wet summers and bitter winters which show up in the various economic and climatic series. In addition, available detailed records from Scotland to Switzerland reinforce the story of climatic extremes. While the wine harvest dates and wheat price indices show only a run of bad years, which do not stand head and shoulders above other bad decades, it is these other series that mark the 1690s as being truly out of the ordinary. Central England Temperatures show this decade as being by far the coldest since at least 1659. All seasons were well below average with the summers being particularly bad. In Switzerland, the story was slightly different. The Pfister thermal index shows the winters and springs as being the coldest on record, whereas the autumns were only on a par with other poor decades, but the summers did not sink to the levels of the sustained cool seasons of the late sixteenth century or the 1810s.

The subsistence crisis started early in the decade in France. Following a second poor harvest in 1693, it brought one of the worst famines since the early Middle Ages.[32] In contrast, in England the impact was much less. Although prices rose, the mortality rates did not respond to these events, and actually fell below trend in the later part of the decade.

The information from other parts of Europe supports these records. In Finland the famine in 1697 is estimated to have killed a third of the population. More generally, throughout Scandinavia the records of the

1690s are littered with reports of crop failures, disasters and abandonment of more marginal land. Moreover, this appears to have been the time when Scandinavian glaciers expanded appreciably, as there is little evidence of expansion before the seventeenth century.[33] The same story emerges from records in Scotland, where between 1693 and 1700 the harvests, principally oats, failed in seven years out of eight in all the upland parishes.[34] Death rates rose to a third to two thirds of the population in many of these parishes, exceeding the figures recorded during the Black Death. The economic consequences of these catastrophic years probably, more than anything, made the union with England in 1707 inevitable.

An explanation of these events is that during the 1690s arctic surface water extended far further south around Iceland (in 1695 the island was entirely surrounded by pack ice) and towards the Faroes. This colder water would have increased the temperature gradient in the southern Norwegian Sea, steering Atlantic depressions on a more southerly course and increasing the incidence of northerly outbursts down into northern Europe and Scandinavia.[35] This would explain the colder winters across the continent. It may also be the reason why Scotland and upland areas of Europe fared so badly in the growing seasons. The delay of spring combined with the high lapse rate in the cold northerly airstreams would have had a greater impact on these areas as such weather conditions usually produced increased cloud with greater rain or snowfall, especially in spring and early summer (Fig. 2.8). This, combined with the fact that they were already operating close to the climatic limit for agriculture, made them particularly susceptible to climatic change. This is a good example of the concept of the non-linear response of agricultural systems to such climatic fluctuations: a concept which was introduced in Section 1.2 and will be explored in more detail in Chapter 6.

As at the beginning of the seventeenth century, the start of the eighteenth century brought considerable relief. With two notable exceptions, the first 40 years of the century were reasonably benign, with the 1730s being particularly mild across all seasons. The first exception was 1709. This was one of those rare occasions when an intensely cold winter killed so much winter wheat, especially in France, that it had a major impact. Although the spring and summer temperatures were average and rainfall plentiful, the Beveridge wheat price index rose to its second highest level above trend in the period 1500 to 1869. The wine harvest date was, however, no later than average. The sharp rise in prices may also have been the product of an over-reaction by both farmers and the market. Severe

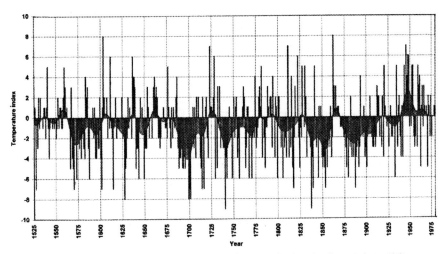

Figure 2.8. The spring temperature index for Switzerland, together with smoothed data showing longer term fluctuations. (Data taken from Pfister, 1995.)

scorching of winter wheat sometimes led farmers to plough up crops, only to see those who had persevered being rewarded by a late but remarkable recovery as the growing season improved. As for the markets, there is evidence that after a number of relatively good years there was a tendency for the market to expect the worst, which stimulated hoarding and rises in prices.

The other interesting incidental feature of the intense cold in France was the impact on the fashion for furniture in England.[36] Because the frost killed two-thirds of the walnut trees in south-east France, which were the source of much of the walnut for English furniture at the time, supplies dried up during the subsequent decade. Forced to look for alternatives, manufacturers turned both to North American walnut and to mahogany from Central America. The latter became fashionable and has remained so since.

The other notable year was the very cold summer of 1725. This year featured the coldest summer (June to August) in the Central England Temperature record. But its impact on wheat prices was small, although the wine harvest date was predictably late. The most striking feature of the 1720s was the mortality crisis in England from 1726 to 1729, which seems to have been unrelated to agricultural or meteorological events. Rather, the cause was a succession of epidemics which were described as 'chincough (whooping cough), Rheumatisms, Inflammations and general scabbiness', but whose precise nature is unknown.[37]

The mild years of 1730s came to an abrupt end with the intense cold of 1740. It was by far the coldest calendar year in the Central England Temperature record. Every month was well below average and it included the second coldest winter (1684 was colder), a very late spring and a cool dry summer. Wheat prices rose sharply and the wine harvest was late. Again there is evidence of over-reaction to the frost damage to winter wheat, which inflated prices. There followed a distinct mortality crisis in both England and on the Continent. But this did not begin in real earnest until late 1741 and it is likely that it was due to infectious diseases, such as dysentery and typhus, rather than to poor harvests.

Throughout the remainder of the eighteenth century the fluctuations in prices and wine harvest dates were less dramatic. Hoskins only ran his analysis up to 1759 because, as he noted, he regarded this declining variability as evidence of improved transportation and increased imports reducing the volatility of prices. Also the increasing cultivation of potatoes provided a buffer in cool wet years when grain yields were low but potato yields tended to be high. In Ireland, however, the habit of not harvesting potatoes, which developed in the mild winters of the 1720s and 1730s, came to an abrupt halt with the great frost of 1740. The fearful mortality that ensued provided a chilling warning of the famine to come a century later, but also ensured that henceforth potatoes were properly stored in clamps or pits for the winter. On the Continent the poor harvests of 1770 and 1771 did, however, cause an upsurge in prices. But the final important subsistence crisis was not until well into the nineteenth century.

Widely known as 'the year without a summer', 1816 is the source of particular fascination to climatologists. Attributed to the massive eruption of the volcano Tambora in Indonesia, the summer was particularly severe in New England, eastern Canada and north-western Europe.[38] The combination of low temperatures, excessive rainfall and unseasonable frosts played havoc with agriculture. Three cold waves ravaged eastern Canada and New England. The first in early June destroyed many crops. A less severe second wave in early July in Quebec and parts of Maine killed crops which had been replanted. The last straw for many was the frosts in the last two weeks of August which killed corn, potatoes, beans and vines in parts of New Hampshire and injured crops as far south as Boston. Although farmers planned for rare late frosts by raising some plants

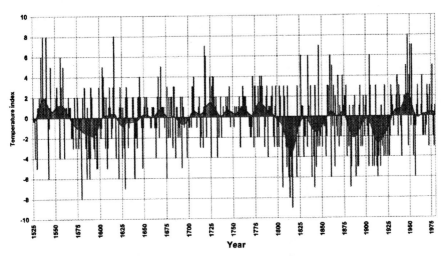

Figure 2.9. The summer temperature index for Switzerland, showing no
significant increase in recent decades together with smoothed data showing
longer term fluctuations. (Data taken from Pfister, 1995.)

indoors to make good damage, the three cold spells in 1816 were unparalleled.

In Europe things were made worse by the disruption of the end of the
Napoleonic Wars and the fact that the 1810s had already featured a series
of cool summers, which pushed the Swiss summer series to its lowest
level since before 1550 (Fig. 2.9). This combination drove the Beveridge
wheat index to its highest level on record, and led to widespread social
unrest, notably in France. The Government was forced to suspend duties
on imported grain and to seek supplies from abroad. In England the
impact was small. In the eastern United States, however, prices soared,
in part, because of the high level of exports to Europe. It also acted as a
catalyst for migration westwards. Emigration was particularly heavy from
Maine and Vermont in the following months. But, in the long run, the
opening of the Erie Canal in 1826 may, however, have been a more
important development in providing farmers with the practical means of
escaping the challenges of the climate and rocky soils of New England.

Even more intriguing is the hypothesis that the disruption in global
weather by Tambora led to the failure of harvests in Bengal in 1816. The
resulting famine triggered a major outbreak of cholera which slowly spread
outwards creating the world's first pandemic of cholera. It reached north-

west Europe and the eastern USA in the summer of 1832. The parallels with the spread of the bubonic plague in the sixth century and the Black Death in the fourteenth century raise interesting questions about links between major volcanoes, bad weather, harvest failures and major pandemics even though the scale of the cholera epidemic was not in the same league as earlier disasters. In England the first outbreak of the disease had only a weak impact on mortality statistics[39] but it had a dreadful impact in Russia, and in New York the death toll exceeded 100 per day.

More generally the picture that emerges from the analysis of specific episodes of poor harvests and high prices is that the weather played its part but was not the dominant factor in high mortality. Statistical analysis by Wrigley & Schofield[40] of the correlation between mortality in England and the Central England Temperature record between 1665 and 1834 shows that mortality increased in cold winters and in hot summers. The main effect of winter cold was immediate but for summer the effect is delayed one or two months. This is consistent: in low temperatures older people died quickly from pneumonia, bronchitis and influenza, whereas in summer younger people died of digestive tract diseases which took longer to kill.

Quantitatively, a one degree of warming in winter reduced annual mortality by about two per cent, whereas a one degree cooling in summer reduced annual mortality by four per cent. The combination of such a mild winter and cool summer was equivalent to raising life expectancy by two years. Wrigley & Schofield went on to note that these generalisations hold for the two subperiods 1665 to 1745 and 1746 to 1834. They note, however, that prices explain a greater proportion of the variance and dominate in the earlier period, but temperature was equally important from 1745 to 1834. Even so, prices only explain about a sixth of the variance.

The challenge that emerges from the analysis is unravelling the complex links between prices, the weather and mortality and the lags and leads between cause and effect. As has already been noted, the relationship between the weather and crop yields is not simple; hence prices will be influenced in a way that will not be revealed by correlating prices with monthly or annual meteorological averages. Moreover, the impact of both weather and prices on mortality depends on the period of analysis. As Wrigley & Schofield observe, the cumulative effect of price variations over five years was essentially zero, as all they did was alter the timing of deaths which would in any case soon have occurred.

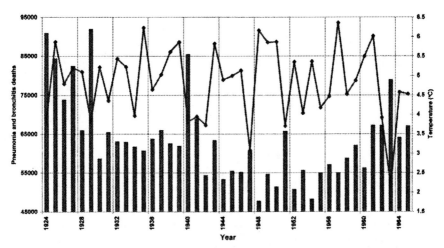

Figure 2.10. The mortality from bronchitis and pneumonia in the UK (vertical bars) compared with the average winter temperature from November to March (solid line), showing the impact of cold winters, and in particular, how the excess mortality declined during the three consecutive cold winters in 1940, 1941 and 1942.

This effect is still of relevance. In the UK where there is a significant excess in winter mortality compared with the rest of the year, the impact of cold winters remains clear in the statistics. But the first cold winter has usually had the most profound impact. A good example of this is the three successive cold winters at the beginning of the Second World War. After a string of relatively mild winters, deaths due to pneumonia and bronchitis in the UK rose dramatically in 1940 (Fig. 2.10). In the next two cold winters the mortality rate due to these diseases fell off sharply so that in 1942 it was below normal, as the most vulnerable had already died. The same phenomenon is likely to occur with other extreme events, such as summer heatwaves, where they come in short order. So predictions of excess mortality due to the increasing incidence of certain extremes need to take account of the fact that the most vulnerable will go first, and so the impact will be less than might be expected on the basis of isolated events.

The longer term fluctuations in the wine harvest and cereal price series also provide useful insights into the nature of the sensitivity of agriculture to climate change. Each example (see Figs. 2.6 and 2.7) have been presented in terms of the annual figures and a smoothed version of the series. The smoothed series is what is known as a 'weighted running-mean' which is designed to remove all the fluctuations shorter than about 10 to 15 years in length.[41] What this shows is that most of the variance in the series occurs at timescales in the range of two to ten years.

If we play more sophisticated games with the series on a computer we can find out more about the frequency properties of the fluctuations. This can be done either by smoothing the series with special numerical filters or by computing a Fourier transform which produces a spectrum of the frequency components of the series.[42] This type of analysis shows there are no dominant cycles in the series (see Section 5.7). There is, however, one interesting difference between the wine and wheat series. The wine harvest data show no evidence of increasing variance at longer periods. This flat response (often termed 'white noise'[42]) reflects the fact that in successive years the harvest date is dependent solely on the weather in that year and not what happened in previous years. In contrast the wheat series show a propensity to increased variance for periods of 10 years and longer. This property is known as 'red noise' and indicates the system has a memory of events over a longer timescale (see Section 5.7). This is not surprising as in the worst harvest failures there is always a temptation to consume seed-corn, thereby reducing the crop in subsequent years. In a run of bad years this effect can permeate the system for a long time and hence increase the longer term variability.

These observations have a number of implications for interpreting the nature of subsistence crises. First, at the simplest level, the fat and lean years came sufficiently close together that the majority of society were able to survive the ups and downs. It was the most vulnerable who went to the wall in all but the most extreme weather-induced breakdowns in food supplies. In extreme events, however, the immediate requirements of survival meant that it was not possible to retain prudent levels of seed for future seasons, the damage done by a bad year was carried forward into subsequent harvests. It often needed a bumper crop to get things back to an even keel. Fortunately, it was in the nature of these interannual fluctuations that good and bad years usually came in close order, and it was only in the extreme cases like Scotland in the 1690s when lasting change can be identified unambiguously.

2.5 Other agricultural crises of the nineteenth century

By switching from subsistence crises to agricultural crises at this point, the scope of the discussion is altered. From around 1820 onwards in

Europe the nature of weather-induced crop failures changed. Improved transport systems reduced the impact of local harvest failures. Also from around the middle of the century the increasing availability of grain imports from Russia, North America and later on from Argentina and Australia changed the nature of the challenges facing societies. Even so, price rises in 1846 and 1847 and the harvest failure in 1848 played a significant part in causing the revolution of 1848 in Germany. Moreover, large-scale imports built up slowly (in Britain at the end of the 1860s 80 per cent of all food was home produced). But as imports increased, the problems of the weather became more and more a challenge for the farming community, rather than society at large.

This shift in emphasis deliberately puts on one side what is without doubt the greatest agricultural disaster of the nineteenth century in western Europe – the Irish Potato Famine. Although the weather played a part in this national disaster in that in both the autumn of 1845 and in 1846 there were periods which were ideal for spreading potato blight (*Phytophthora infestans*), the weather was not the principal cause of the famine. Other factors, including the emergence of a new virulent pathogen, high population density and the reliance on a monoculture, played a much more important role in the calamity. So, in terms of the issues under consideration here, it is not a good example of how the weather can disrupt economic systems.

Of more direct relevance is the events of the late 1870s in Britain. In spite of the increasingly dominant role of industry in the economy in the last quarter of the nineteenth century, agriculture constituted some 10 per cent of gross capital formation. Although this proportion was declining, marked fluctuations in this sector had an important influence on the overall performance of the economy.[43] The biggest fluctuations occurred in the late 1870s, which were characterised by a series of exceptionally wet years. These hit British agriculture at a time when it was facing increasing competition from imports of grain.

The excessively wet year of 1877 brought a substantial drop in production, but pride of place has to go to 1879. This year began and ended with exceptionally cold winters and in between had an unrelentingly cold wet growing season. The Central England Temperature record shows that 1879 was the coldest year in the last two and a half centuries. Every month was below average, and together with November and December 1878 and January 1880 there were 15 consecutive cold months. In south-east England the incessant rains produced saturated ground with the highest summer soil moisture figures since the late seventeenth century.[44] The

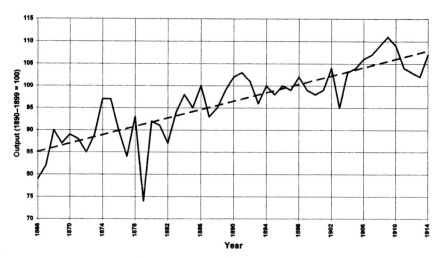

Figure 2.11. Index of British agricultural production in the late nineteenth century showing, the impact of cold wet year 1879. The dashed line shows the linear trend for the period. (Data from Phelps-Brown & Hopkins, 1981.)

impact on British agricultural production was dramatic (Fig. 2.11). The reduction below trend in total output was around 20 per cent, while figures for the six principal crops in Ireland showed a nearly 35 per cent fall below trend.

The longer term economic consequences of this shock to the agricultural system were, however, complicated. Solomou,[45] in analysing how variations in agricultural output contributed to long term economic cycles (see Section 6.3), noted that supply shocks in the 1870s led to increased imports of agricultural commodities but did not lead to a fundamental adjustment in the economy. Cheap agricultural products were not flooding the British and European markets. Increased imports were purchased at relatively high cost. During the 1880s, international competition increased but, in Britain, agriculture experienced a revival of output and productivity only to suffer a setback in the 1890s, in part as a consequence of a series of warm, dry summers. Overall, climatic swings were an important factor in the agricultural swings in the second half of the nineteenth century. But the agricultural sector only played a significant part in the cycles experienced by the British economy in the 1870s and 1880s. Thereafter, the impact of climatic change on agriculture did not have a measurable impact on the economy, which suggests that technical progress was making the system less vulnerable to climatic shocks.

2.6 The Dust Bowl years

To complete the historical examples of the agricultural impact of extreme weather, the Dust Bowl years of the 1930s in the Great Plains of the Midwest USA provide a fitting climax. Although the images of the catastrophic effect of drought of those years are commonplace, it is easy to overlook the enormity of the meteorological events that struck the USA in 1934 and 1936.

In part, this jaded reaction reflects the extravagant nature of the climate of much of North America. This part of the world has more than its fair share of extreme weather. Not content with fearsomely cold but erratic winters and stifling summers, the continent experiences some of the severest thunderstorms and more tornadoes than almost anywhere else in the world, as well as some of the worst hurricanes. As a result, its recorded history is littered with weather disasters. These extremes are particularly relevant to the Great Plains, where settlers froze to death in blizzards and bitter winters, saw their townships obliterated by tornadoes, or were forced to abandon settlements when drought destroyed their crops.

The vulnerability of early European settlers to the extremes of Great Plains weather makes it easy to fall into the trap of underestimating the Dust Bowl years. Furthermore much of the damage was attributed to unwise farming practices, and subsequent action to prevent a repetition of the damage succeeded to a considerable extent, so it is easy to overlook just how abnormal the weather was. This could be short-sighted given the fuss that has been made about the US droughts and heatwaves of 1980 and 1988 (see Section 4.4). When viewed alongside 1934 and 1936, these more recent events which at the time were widely seen as evidence of the Greenhouse Effect in operation, look less impressive.

The winters of the Midwest USA are cold and relatively dry. The summers are hot (average temperatures in Kansas for July and August are over 27 °C with daytime highs averaging 34 °C), which means that crops lose a lot of moisture through the process of evapotranspiration. This need not be a problem as the summers deliver much of the annual rainfall. The fluctuations from year to year are, however, large. Moreover, the hot summers are the drought years, so when the rains fail, agriculture is in double jeopardy as the crops wither rapidly in the blazing heat. The combination shows up clearly in Fig. 2.12, and the extreme conditions of 1934 and 1936 stand out dramatically. These two summers were by far the hottest since 1900, and 1936 was the driest by a large margin; 1934,

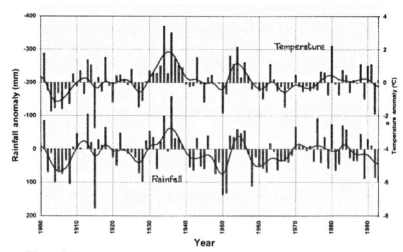

Figure 2.12. Summer rainfall and temperature figures for Kansas during the twentieth century, showing how the heat and drought of 1934 and 1936 stands out and the close correlation between hot and dry summers which shows up best in the smoothed data. (Data from National Climatic Data Center, Ashville, N.C., USA.)

although hotter, was just edged out of second place by 1913. Both were preceded by exceptionally dry springs. More generally the 1930s drought has variously been calculated to have been a one in 250 to one in 400 year event, although tree-ring studies suggest that comparable widespread drought occurred in the 1790s, 1820s and 1860s.[46] What is certain is that 1934 and 1936 put subsequent summers in the shade; 1980 came in third place, being slightly hotter and drier than 1954 in Kansas, while 1988 can only claim to be in the top 20 per cent of the hottest, driest summers in this part of the world.

These figures provide only a partial picture of the extent of the drought during the 1930s. A more comprehensive measure of drought is an index which was developed by Wayne C. Palmer in the 1960s.[47] Now used to produce a weekly analysis of the soil moisture condition across the USA, this index used weekly precipitation statistics and mean temperature to compute evapotranspiration deficits or surpluses, which are compared with previous weeks and climatological averages to derive a crop-moisture index. This is expressed in a scale ranging from +4 or above (extremely moist) through zero (normal) to −4 or below (extreme drought). When calculated retrospectively, the maps for 1934 and 1936 show that the area of severe drought was far more extensive and prolonged than in subsequent drought years such as 1980. The analysis of past US droughts has also been extended back in time using tree-ring analysis.[48] This work confirms the enormity of the Dust Bowl years, with no year back to 1700 matching 1934, although periods of bad years around the middle of the eighteenth century exhibited sustained dry conditions comparable to the 1930s. The other interesting feature of this work is the substantial evidence of a 20-year cycle in the incidence of drought in the USA west of the Mississippi. Potentially more important is a recent analysis of lake sediments in North Dakota[49] which indicates that prior to 1200 AD there was a greater frequency and intensity of droughts in the northern Great Plains of the USA. This analysis extends back 2300 years and shows that periods of extreme drought lasted for centuries. It also has striking evidence of a cycle of around 18.5 years. So any interpretation of current rainfall fluctuations in the Midwest of the USA has to take full account of these past ups and downs.

The impact on agriculture of the Dust Bowl Years can be measured in terms of both the drop in productivity compared with trend values, and in the scale of abandonment. In 1934 and 1936 the average wheat yields across the Great Plains fell by about 29 per cent compared with the trend. Moreover, much of this loss related to the fact that nearly 40 per cent of the land sown was not harvested, compared with typical values of around 10 per cent. When it comes to abandonment a more interesting picture emerges. From the hardest hit parts of Kansas and Oklahoma there was massive outward migration, with more than half the population leaving. But the intervention by the Democrat Government – which had been swept to power at the end of 1932 as a reaction to the laissez-faire economics of the Republican Party and its failure to address the problems of the Great Depression – did alleviate the worst problems. By comparison, the flight of settlers from the Great Plains during the far less extreme drought in the 1890s was much greater.[50] This was because at the end of the nineteenth century the agricultural frontier was still only poorly integrated into the national economy and the farmers were ill equipped to handle drought. The immediate consequence for much of the Great Plains was widespread massive depopulation. So the first message from the Dust Bowl years is that government intervention can have a substantial impact in alleviating the consequences of extreme weather events. As part of the

Figure 2.13. An example of the damage caused by drifting sand in Texas during the Dust Bowl years. (Reproduced by permission of Hulton Getty.)

New Deal, the provision of disaster aid enabled many people to survive the drought and set the precedent for similar action whenever severe weather strikes in the USA.

The more fundamental lesson that emerged from government studies at the time was that much of the agriculture on the Great Plains was not appropriate to such an arid region. In the wetter years of the 1920s there had been a great expansion on to marginal land. The light sandy soil and low rainfall was not suitable for arable farming and should have been left as pasture. Without grass to hold the soil in place, fierce winds and scorching heat stripped off huge quantities of topsoil in the frequent duststorms – the terrifying 'Black Rollers' – that made life in the Midwest well-nigh unbearable (Fig. 2.13) and blotted out the Sun 3000 km away in the cities of the East Coast. Indeed, the greatest cost of the Dust Bowl years may have been the destruction of agricultural land by the complete removal of topsoil.

The longer term response was to purchase marginal land, retire it from cultivation and seed it with grass. This was combined with educational programmes for farmers to plant trees for shelter-belts, grow crops better equipped for drought conditions, and introduce conservation methods (e.g. contour ploughing, water conservation in irrigation ponds and strip ploughing to allow part of the land to lie fallow). The advantage of increasing rainfall meant that the lot of Midwest farmers improved into the 1940s. But, as a result of the demands of food production during the Second World War, the temptation to press marginal land back into production proved irresistible. While the worst abuses of the 1930s were not repeated, the drought of the early 1950s was nearly as severe, so many farmers had to suffer the experiences of the past. This reminder did, however, reinforce the case for the state and federal laws to protect the land from overexploitation.

A number of threads can be drawn together from this final historical example of the agricultural disruption caused by extreme weather. These consist of a combination of generic problems, which occur in every case, and the sense of progress in the developed world which increasingly has enabled societies to ride out the worst of the weather. The vulnerability remains unchanged; the overexploitation of marginal land, often associated with population pressures, makes communities far more sensitive to even the normal fluctuations of the climate. When fluctuations reach around the one-in-a-century level then complete breakdown occurs and outward migration is often the only option available to many people. The 'Okies' driving with all their possessions from Oklahoma to California in 1936 were taking the same desperate course as the 'Sea Peoples' may have adopted in the thirteenth century BC. The reception they received when they reached California was, however, marginally less hostile than meted out by Ramesses III to the northern invaders. Throughout time this escape strategy has been the avenue of last resort and the hostility is an equally desperate response. Only in the case of the Norse colonists is there no evidence of their trying to escape. If they did take this course, it too failed.

The sense of progress is seen both in the response of central government and in the growing awareness of how to respond to extreme weather. When drought returned to the Great Plains in the 1950s, the combination of the existing state and federal programmes and the adoption of agricultural practices to accommodate dry conditions prevented outward migration. So any exodus was negligible compared with the 1930s or the almost total depopulation of the 1890s. The combination of government fiscal and legislative action has to be seen as an important factor in this more resilient response to drought.

This final example can be regarded as the end of the beginning of our analysis of the economic effect of climatic change. Henceforth, in the developed world, the involvement of central government means that we are concerned with national and sometimes international response to extreme events. Furthermore, while there has been a whole series of mass-ive weather-related disasters in the developing world since the Second World War, increasingly the response has involved international inter-vention and attempts to co-ordinate national and international efforts. So from now on we will be looking in more depth at an integrated reaction to both the meteorology and its consequences, and what this may mean for the future climatic change. In one sense this could be a case of moving on to the beginning of the end, in that in the future there will be no escape strategy for an overpopulated world. Whatever the climate does we will have to come to terms with the changes and rise to the challenge wherever it arises.

2.7 Notes

1 Chapter 1 of Wigley, Ingram & Farmer (1981) provides an illuminating review of these issues.

2 Lamb (1995), p. 131.

3 Sandars (1985) provides an interesting analysis of the possible origins of the 'Sea Peoples'. While not subscribing to the climatic theories for their movements, this book provides fascinating background information to the events of the period.

4 Kennedy (1986).

5 Mango (1988), p. 71.

6 Ibid., p. 73.

7 Carpenter (1966). This book argues that the decline in Bronze Age civilisations in the eastern Mediterranean also had climatic origins.

8 The whole question of the cataclysmic event around 536 AD is the subject of intense debate. Stothers (1984) identified it as the eruption of Rabaul, but there are alternative explanations, which are reviewed in a compelling manner in Chapter 6 of Baillie (1995). This book also identifies three comparable earlier events. The first appears to have been the eruption of Thera, on the island of Santorini, in the Aegean Sea, around 1627 BC, which may have led to the collapse of the Minoan civilisation on Crete. The second was in 1159 BC, which Baillie links with the onset of the 'dark

age' in the eastern Mediterranean, and the third is a more shadowy event in 207 BC.

9 Hodell, Curtis & Brenner (1995).

10 The water in the lake evaporates rapidly in the dry season, with water containing the lighter oxygen isotope (^{16}O) being more likely to evaporate than that containing the heavier isotope (^{18}O). So the ratio of these two isotopes is an indication of balance between evaporation and precipitation in the catchment area at any point in the past. Shellfish in the lake used oxygen in the water to manufacture calcium carbonate in their shells. Measuring the isotope ratio in the shells laid down in the sediment provides a detailed picture of the dryness of the local climate over the years.

11 Thompson (1993).

12 Dansgaard *et al.* (1975).

13 See Chapter 17 by T. H. McGovern in Wigley, Ingram & Farmer (1981).

14 The broad flavour of the apocalyptic events of the fourteenth century is captured in Tuchman (1978).

15 Lamb (1995), p. 191

16 Kates *et al.* (1985), p. 365.

17 Hammer, Clausen & Dansgaard (1980).

18 Lucas (1930).

19 Lawrence (1972).

20 Phelps-Brown & Hopkins (1956)

21 Montieth (1981).

22 Lamb (1995), p. 197.

23 Phelps-Brown & Hopkins (1956).

24 Le Roy Ladurie & Baulant (1980).

25 Beveridge (1921).

26 Hoskins (1964) and (1968).

27 Wrigley & Schofield (1989).

28 Pfister's work is found in its most accessible form in Bradley & Jones (1995), Chapter 6.

29 Manley (1974).

30 Parker, Legg & Folland (1992).

31 Van den Dool *et al.* (1978).

32 Le Roy Ladurie (1972).

33 Grove (1988), Chapter 3.

34 Lamb (1995), p. 222.

35 Ibid., p. 217.

36 Burroughs (1982).

37 Wrigley & Schofield (1989), p. 664.

38 Stommel & Stommel (1979).

39 Wrigley & Schofield (1989), p. 655

40 Ibid., p. 389.

41 Burroughs (1994), p. 171.

42 Ibid., see Appendix 1 for a general discussion of the analysis of time series.

43 Solomou (1990), p. 122.

44 Wigley & Atkinson (1977).

45 Solomou (1990), p. 123.

46 See Chapter 16 by D. M. Meko in Bradley & Jones (1995).

47 Oliver (1981), p. 129.

48 Mitchell, Stockton & Meko (1979).

49 Laird *et al.* (1996).

50 See Chapter 21 by M. J. Bowden *et al* in Wigley, Ingram & Farmer (1981).

3

Cold winters

'Unimportant, of course, I meant,' the King hastily
said, and went on to himself in an undertone,
'important – unimportant – unimportant – important –'
as if he were trying which word sounded best.
Alice's Adventures in Wonderland, Chapter 12

So far cold winters have only received passing mention. As the vulnerability of Europe to weather-induced subsistence crises declined with improved transport systems, increasing access to overseas food supplies and growing affluence, the impact of cold winters assumed greater importance. The response of many aspects of industrialised societies to the disruption caused by snow, ice and prolonged cold is non-linear. This is because the degree of disruption rises disproportionately as the temperature falls further below normal. The consequences of severe winters have been recognised in the underlying mortality statistics and in certain particularly poor harvests (e.g. 1709). Furthermore, the freezing of rivers and canals interrupted the production and distribution of food. In medieval England when there were up to 6000 water mills for grinding flour, the freezing rivers could stop the production of bread and so lead to severe shortages. But, by comparison with the disruption of more industrialised societies, these problems seem relatively minor.

There is also a statistical reason for considering the consequences of cold winters. This is that in mid-latitudes, especially in the northern hemisphere, the variability in winter weather is greater than in other seasons. Whether measured from year to year, from decade to decade, or as long-term trends, the biggest changes are seen in winter. The fact that these fluctuations are amplified in terms of the incidence of snow and ice reinforces the non-linear response of industrial societies to severe winter weather and highlights how climatic change alters how we manage our lives.

As in Chapter 2 this analysis will focus initially on European examples, although much of North America experiences colder winters than western Europe. The reasons are twofold. First, the longer meteorological records make it easier to put recent extremes in context. The second is a recognition of the variable nature of winters in North America, especially in the eastern half of the continent. It is a feature of the climate of the region that even in relatively mild winters it will experience at least one savage cold wave, often accompanied by penetrating winds and heavy snowfall. While these cold snaps may be short lived, they ensure much of the country remains accustomed to such extremes. In colder winters, these cold waves become more frequent and sustained, but not necessarily more intense. In contrast, north-west Europe, and especially the UK, may have a series of winters without any appreciable snowfall and freezing temperatures. As a consequence, after a run of mild winters a severe snap causes much greater disruption. So the UK provides some of the most illuminating examples of the chaos caused by abnormally cold winter weather. But, as we will see, the English experience is of wider relevance, as the USA demonstrated in the late 1970s.

3.1 Second World War

Before embarking upon the disruption caused by cold winters to economic activity, there is a more striking example of how unusually cold weather can intervene in history. This is the three cold winters at the beginning of the Second World War. In terms of meteorological statistics, in central northern Europe there had only been one very cold winter (defined as falling in the lowest 10 per cent) since 1895. This was 1929. So, to have three consecutive winters which achieved this mark (1940, 1941 and 1942) was not only statistically notable, but also unparalleled in the previous two centuries. It was, however, the implications of this string of severe winters for the development of the war that matters most.

The intense cold of the first winter of the war is often overlooked. Winston Churchill's *History of the Second World War* makes no mention of it, but Field Marshall Viscount Alanbrooke, who, as General Alan Brooke, had commanded the British forces in France in 1940 and subsequently became Chief of Imperial General Staff, was later to conclude that the weather saved the British Army.[1] The meteorological facts are that a cold wet October was followed by a mild but exceptionally wet November. Then in December the severe weather set in, heralding the coldest winter on the Continent since 1830. For much of December, January and February, northern France and Belgium were covered by deep, powdery snow, and the roads were sheets of ice.

The effect of the weather on military operations was profound. On 23 November, Hitler announced his intention to attack the Anglo–French at the earliest possible moment. But throughout December the weather was never good enough to make full use of the *Luftwaffe*. On 10 January the invasion of the Low Countries was ordered by Hitler for 17 January. But the plans fell into the hands of the Belgians, and while the Germans were considering the implications of the breach of security, the heavy snow began to fall again and the offensive was cancelled. The cold weather ran well into February and ensured hostilities were delayed until the spring.

Alanbrooke observed that the extremity of the bitter winter prevented Hitler from launching an attack against an ill-equipped and ill-prepared Anglo–French army. Throughout the cold weather the British Expeditionary Force laboured ceaselessly to build up its defences and by the spring had doubled its numbers. Given the way this army was swept aside in May 1940, it is easy to see why Alanbrooke concluded that, but for the cold winter, the Germans would have been at the Channel ports several months earlier when the RAF had 20 fewer squadrons than during the Battle of Britain. No wonder he said 'What might have happened if the Germans had attacked before the winter, I shudder to think'. Another indicator is how, as a last desperate throw, Hitler launched the Ardennes offensive during the winter of 1944–45, which happened to coincide with a cold spell. The relative success of this offensive without command of the air against an exceedingly well-equipped and well-trained Anglo–American army provides further evidence of what might have happened if the weather had not been so severe in early 1940.

An alternative view of the events in January 1940 is taken in Liddell Hart's history of the Second World War.[2] He argues that, while the weather did play a part, the loss of the German plans was a more significant factor. It led to a complete recasting of their plans. Instead of striking through central Belgium, they adopted a much bolder approach, proposed by General Manstein, of the mass of German tanks driving through the Ardennes. This offensive in May took the Allies by surprise and laid

France open to defeat. The earlier plan, if executed during the winter, might have succeeded in the Germans reaching the Channel ports, but by taking the Anglo-French forces head on, it might well have bogged down along the Somme.

So, if the course of the Phoney War seems too speculative, then the events of the winter of 1941–42 offer even more food for thought. In between, the winter of 1940–41 had been very cold, but it exerted a less appreciable influence on military campaigns, although the cold, late, wet spring in eastern Europe was a factor in delaying the start of the German campaign against Russia. Late 1941 was a different matter. During the summer, by switching objectives, Hitler had probably missed the opportunity of early victory in Russia. By early October everything depended on Operation Typhoon – the attack on Moscow. The Germans retained enough strength to defeat the Soviet armies, providing the weather remained mild enough not to interfere with the offensive.

The Germans were not ignorant of the severity of Russian winters. The reputation of 'General Winter' was well recognised, even though Napoleon's defeat in 1812 had little to do with abnormally cold weather until the final stages of the retreat from Moscow. What the Germans were relying on was their assumed expertise in predicting the weather. Under Franz Baur, the German weather service had ostensibly developed a considerable expertise in seasonal forecasts. Baur, who had the ear of Hitler, confidently told the military planners that the winter would be mild. This was based on a set of rules built up from climatic records. It was also influenced by the statistical improbability of having three very cold winters in a row. But the facts denied the statistics. In European Russia, 1941–42 broke all records. From early November, repeated bouts of Arctic air flooded the country. Around Moscow the five-month period from November to March was probably the coldest in at least 250 years and played havoc with the German offensive.[3] By the end of November when an average temperature range between 0 and −10 °C might be expected, minima as low as −40 °C were recorded. Despite these facts, Baur remained so confident that his forecast was accurate, at first he refused to believe the weather records sent back to Berlin.

The consequences for military operations were dramatic. Below −20 °C there were widespread weapon and machinery malfunctions. Firing pins shattered, hydraulics and lubricants froze, rifles, machine guns and artillery failed, and tanks and supply trains ground to a halt. Many troops were incapacitated by frostbite because the intense cold arrived before supplies of warmer clothes reached the front. It is estimated that between 100 000 and 110 000 German troops were lost through frost-related deaths between 4 October 1941 and 30 April 1942.[4] During the same period some 155 000 died or went missing as a result of enemy action, and in the depths of winter it is probable that frostbite casualties exceeded those resulting from enemy action.

It remains a matter of debate among military historians as to whether the bitter cold was the decisive factor in the failure of Operation Typhoon. What is clear is that the German forces could have operated more effectively had the weather been average or milder than normal. It also confirms the vital importance of being able to forecast longer term fluctuations in the weather. As we will see later, modern meteorologists continue to experience the same trials and tribulations that afflicted Franz Baur in 1941.

More generally, the cold winters of the early 1940s provide good examples of fateful events. Whatever their actual consequences, there can be no doubt that things would have followed a different course if the weather had been more normal. This does not mean, however, that the outcome would have been radically altered. The forces that eventually settled the Second World War might still have prevailed. It is not the purpose of this book to speculate on the alternative courses of history. Suffice it to say that cold winters intervened in the course of the war in two notable instances, and in both cases they served to frustrate German plans.

3.2 The 1947 fuel crisis

As if three cold winters were not enough, the 1940s had one more trick up its sleeve. This was the winter of 1947 which gripped much of Europe and reduced the southern half of Britain to a state of economic paralysis for much of the month of February. As such it provides an object lesson of how the combination of extreme weather and other economic circumstances that can dislocate a modern society. This particular mixture of weather conditions, underinvestment in infrastructure and breakdown in supplies of essential services provides a vivid example of the challenge of abnormal weather. While many features of Britain in the early post-war period were exceptional, the underlying messages of the fuel crisis are still relevant today.

Before discussing the weather of February 1947, it helps to set the political scene. The Labour Government, elected in the summer of 1945, had embarked upon a massive programme of nationalisation. This included the coal mining industry, the electricity supply industry and the railways. All these essential services had emerged from the War desperately short of investment and were badly overstretched to meet the needs of an economy that was expanding rapidly to meet the peacetime needs of both the domestic and international markets. So it was well understood that any disruption of basic services would have a major impact on the economy. Desperate efforts were being made to increase the capacity of these services as they came into public ownership. But in the case of electricity generating capacity this depended on investment decisions that had been made several years earlier.

It is, however, a measure of the problems facing the energy industry that public concern about coal supplies led to many people purchasing electric fires to offset coal shortages. Monthly sales of electric fires in 1946 ran at about the same rate as the total annual increase in the capacity of the electricity supply industry. So potential demand was rising by an order of magnitude faster than the industry's capacity to meet it. The increasingly frequent power cuts in late 1946, however, were a clear sign that this stratagem was flawed. Harold Hobson, the Chairman of the Central Electricity Board wrote to the Minister of Fuel and Power, Emmanuel Shinwell, on 12 December 1946 saying the Board had underestimated demand that 'the most significant factor was the unrestricted sale of electric fires and immersion heaters – the substitution of electric fires for coal had increased demand by 10 per cent'.[5] So the predictable but perverse public response to the potential crisis was to take action which could only amplify the problem.

In these difficult circumstances the coal mines came into public ownership on 1 January 1947 as the flagship of the Government's nationalisation policy. At the time the winter was not particularly severe, and for nearly three weeks it seemed that the Government might squeeze through with the slender coal stocks and inadequate generating capacity at its disposal. This would have been difficult to achieve even in the most clement weather as power station stocks were down to four weeks supply by the beginning of the winter, compared with the standard maintained before the war of 10 to 12 weeks. So any disruption in coal supplies or increase in demand would cause problems as a minimum of two weeks' stocks was

Figure 3.1. Railwaymen clearing snow on the Settle-Carlisle line in northern England during February 1947. (Reproduced by permission of the National Railway Museum.)

needed to keep the system operating, given that amounts varied appreciably from station to station.

As it was, the weather was in a malign mood. Around 22 January it turned dramatically colder, and was destined to remain so until mid-March. Persistent high pressure formed close to northern Britain and continuous bitter easterly winds covered much of the country. Worse still, the conditions brought frequent heavy snowfall to many areas (Fig. 3.1). Although the night-time temperatures were not exceptionally low, almost continual cloud, biting winds and below freezing daytime temperatures meant that February 1947 was the coldest February in the Central England Temperature record stretching back to 1659 (see Section 2.4).

These conditions were combined with exceptional snowfall, especially in upland areas, which trapped flocks and paralysed transport. Generally, it was reckoned to be the snowiest winter since 1814. Worse still, when following massive snowstorms across the north of the country in the first half of March, the thaw eventually arrived with heavy rain, making it the wettest March in at least 250 years. The combination of melting snow and incessant rain produced record-breaking floods. By the end of March some 700 000 acres of farmland were under water.

The impact was immediate and dramatic. On 29 January the potential demand of just over 11 gigawatt (GW) exceeded the maximum grid output of 9.27 GW by 16 per cent. There were power cuts of up to 12 hours, load was shed and the frequency of the supply reduced. An immediate minor problem was that all the electric clocks in the country started running slow. By Wednesday 5 February, Harold Hobson had to inform Emmanuel Shinwell that coal supplies to London and the south-east had ceased, as colliers could not sail from the north-east coalfields because of the easterly gales. As a consequence, all supplies of electricity could cease by the end of the week unless stringent controls were introduced. So on Friday 7 February Emmanuel Shinwell announced to a dumbstruck Parliament that from 10 February all electricity supplies to industry would be cut, except where it was needed to protect plant. Domestic and commercial supplies would be cut between 9 and 12 a.m. and from 2 to 4 p.m.

These restrictions did not stop the demand for electricity but they gave the industry breathing space. By eking out stocks, and with heroic efforts to move coal from the pits to the power stations, the country managed to get through the worst of the cold spell. Supplies to industry in central England were restored on 24 February and in the north-west and south-east of the country on 3 March, but voltage reductions continued until 30 March. Restrictions on domestic consumers were lifted on 4 May, but by then many people were getting round them. But a ban was imposed on electricity and gas (produced from coal in those days) use for space heating. This final restriction was hardly necessary as, by a perverse meteorological coincidence, the country then had the hottest May to September period in the last three centuries.

The immediate consequences of the cold weather were the immense damage to agriculture and the disruption of industrial production. In the case of agriculture the immediate impact was that some four million sheep died. The overall mortality rate was around 20 per cent, but in some hill farming areas it was as high as 90 per cent. Combined with the floods of March, British agriculture was in a dire state as spring emerged from the ruins of winter. The Government set up a disaster fund.

The impact on British industry was a sharp drop in industrial pro-

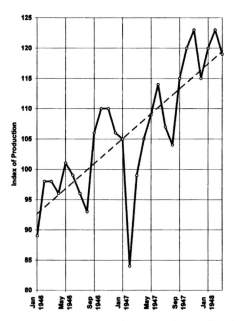

Figure 3.2. Decline in UK industrial production during the winter of 1947.

duction (Fig. 3.2) and a nearly tenfold increase in temporary unemployment to nearly two million. While things soon returned to normal, the underlying damage was profound. It came in two forms. The first was the impact on the credibility of the Labour Government. While many of the changes it introduced took root and became a part of national life, the failure to handle the fuel crisis sowed the seeds of doubt in the electorate's mind about its ability to manage the economy and the benefits of nationalisation. More directly it accelerated the economic problems of the country; the loss of £200 million in exports contributed to a record balance of payments deficit in 1947 (£630 million – a huge figure those days) which ushered in the devaluation of 1948 (sterling was devalued from $4 to $2.80) and the period of austerity with Sir Stafford Cripps' belt-tightening budget of that year.

Even more subtle was the effect the crisis had on the UK's standing with the USA. In his book *Cold Winter: Cold War*, Robert G. Kaiser[6] argues that the dire straits of the UK in February 1947 were instrumental in the US decision to provide aid to Greece and Turkey, which led to the Truman Doctrine and ultimately to the Cold War. Until the UK's economic collapse, the USA had assumed that Britain would take the lead in the eastern Mediterranean, and share the burden of the occupation and defence of Europe. At a time when the Government was also grappling with the issues of withdrawing from India and Palestine, it is easy to see why the US Administration concluded the UK had lost the capacity to maintain an independent role as a world power. Furthermore, much of the rest of Europe was suffering equally badly from the extreme winter, so the USA felt compelled to intervene to prevent a complete collapse.

To the extent that the winter led to a fundamental change in US policy, and with it the Marshall Plan, the more active involvement of the USA in the defence of Western Europe, and the eventual establishment of the European Economic Community, this must be regarded as the most lasting impact of the fuel crisis. The USA confronted by the twin challenges of the restless hostility of the Soviet Union and the ruined economic condition of Europe, took on an international role which was to dominate global diplomacy until the collapse of the Soviet Union at the end of the 1980s. It also adds weight to the emerging conclusion about the non-linear response of society to extreme weather events: the nearer the edge a system is the greater and more lasting the consequences of being temporarily tipped into chaos.

3.3 The winter of 1963

The sustained cold spell of January and February 1963 broke many records in northern Europe. In central and southern England it was the coldest winter since 1740 (Fig. 3.3), while across Europe it was the coldest since 1830. It also featured exceptionally cold weather across the USA and in Japan, and it is often cited as the classic example of meridional global weather patterns which bring extreme winter weather in the Northern Hemisphere (see Section 5.7). In every respect but one it was more extreme than 1947 – in the UK it did not feature the frequent snowfall that was such a disruptive factor of the earlier fuel crisis. Nevertheless, it provides an interesting comparison with the events of 1947.

In the UK, following the worst bout of freezing fog in London since the 'Great Smog' of December 1952, the cold weather started in earnest on 23 December 1962. Heavy snowfalls on Boxing Day and then on 29 and 30 December covered much of the country. In many parts of rural lowland, central and southern England, this snow lay until the beginning of March. Throughout this period there was sustained cold with periods

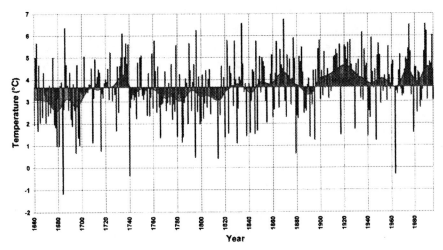

Figure 3.3. The Central England Temperature record for winter (December to February) showing that the 1962–63 winter was the coldest since 1740. This series, together with smoothed data showing longer term fluctuations, also shows how winter temperatures in England have risen in the last two centuries. (Data from Manley, 1974, and Parker *et al.*, 1992.)

of more intense frost and freezing fog and occasional falls of light snow. The west of the country, and notably Northern Ireland, had a further massive snowfall at the beginning of February. Unlike 1947, the winter of 1963 did not end with a bang but slowly faded away as temperatures returned to normal in March.

The sustained effects of the cold weather, apart from disruption of road and rail traffic, first appeared with widespread power cuts on 4 January. Although matters were made worse by a 'work-to-rule' called by the Electrical Trades Union, who were engaged in a wage dispute with the Electricity Council, the huge demand triggered by the cold was the real culprit. The scale of disruption of both electricity and gas supplies increased throughout January. These came to a head on 24 January when the maximum potential demand was estimated to be 32.1 GW, a rise of 15.2 percent on the figure in the previous year. The Central Electricity Generating Board (CEGB) could only meet 29.52 GW.[7] Then the build-up of polluted hoar frost on the insulators of the National Grid started to conduct electricity. This resulted to a huge number of 'flashovers' and a virtual breakdown of the distribution network.

The economic consequences of this winter in the UK and Europe were far less damaging than in 1947. In spite of the extreme cold, the absence of any basic supply constraints and the buoyant state of the economy minimised the impact of the winter. The seasonally adjusted index of total industrial production in January 1963 was 7 per cent below trend, while the adjacent two months showed half this drop in output. This decline amounted to £300 to £400 million lost output. The construction industry was particularly hard hit. The number of housing starts and completion fell by more than 40 per cent during the first quarter of 1963. The value of construction work in this period was £140 million below that of adjacent periods. The seasonally adjusted total inland fuel consumption in the first quarter of 1963 was 7.4 per cent above trend, despite the fall in industrial output. The seasonally adjusted index of consumption of gas, electricity and water, rose 13 per cent in January and 15 per cent in February. But, as has been made clear, this supply failed to meet demand in spite of the heroic efforts of the energy industries.

A comparable analysis of the figures in the UN Monthly Bulletin of Statistics for other European countries in 1963 produces a similar, though less dramatic story. Seasonally adjusted figures for industrial production in Poland and West Germany show about a 5 per cent drop in January and February 1963. In West Germany the seasonally adjusted unemployment rose by about 120 000 during those two months. Somewhat surprisingly for countries better equipped to handle the effects of anomalously cold weather, the indices for production in the construction industry showed marked declines. Most notably, output in West Germany dropped by over 50 per cent, on a seasonally adjusted basis, while in France the corresponding drop was about one-third.

By comparison with 1947, both the immediate and lasting consequences of this extreme weather were small. This largely reflects the robust economic state of the UK, and for that matter other European countries. But the hidden costs are no less interesting. Because of the heavy reliance of the UK on electrical storage heaters in the early 1960s, before the halcyon days of North Sea gas, the political consequences of failing to meet domestic demand were considerable. It so happened that the Parliamentary Select Committee on Nationalised Industries was examining the electricity supply industry during the winter. On 23 January Sir Christopher Hinton, the Chairman of the CEGB, was giving evidence and was closely cross-questioned on the problems of meeting public demand. He explained the difficulties experienced as a result of the extreme demand, the shortage of other fuels, the limitations of the National Grid and industrial action. He defended the industry's forecasts and the use of the figure of a 14 per cent planning margin – the additional plant needed over and above predicted peak demand to enable the industry to meet extreme demand due to cold weather and to cover plant breakdowns.

The committee, influenced by the events of the winter, was not convinced. In its report published later in the year it concluded that the planning margin was too low and saw no good reason why security of supply should be inferior to that in the USA, Canada and France. It noted that the industry's forecasts had been low and recommended that the electricity supply industry should aim to achieve a security of supply at least equal to that enjoyed by other advanced countries. This report combined with the events of the winter, led to the planning margin being increased to 20 per cent.

At the same time, the forecasts of economic growth took a particularly positive line. The net result was that the CEGB's forecasts for peak demand (simultaneous maximum demand in an average cold spell – SMD) six years ahead (the time needed in the mid-1960s to build new power stations) shot up. The Board predicted in July 1963 that the SMD would grow by 7.9 per cent per annum over the next seven years, with peak demand reaching 54 GW in the winter of 1969–70.[8] In practice this value had not been reached by the end of the 1980s when the industry was privatised (Fig. 3.4), by which time the UK had sustained 20 years of excessive generating capacity. Although this capacity, especially the oil-fired plant, was a major factor in the defeat of the Miners' Strike in 1984, the cost of this overinvestment has to be reckoned in billions of pounds sterling. It can be argued that the discovery of huge reserves of natural gas in the North Sea in 1965, and its subsequent displacement of electricity from the domestic heating market, could not have been anticipated. Nevertheless, the catalyst for the series of decisions to increase generating capacity was the inability to meet the huge surge in demand as people tried to keep warm during the intense cold of January 1963.

3.4 The winter of discontent

Any discussion of the impact of cold winters on Britain must include the events of January and February 1979. Widely known as the 'winter of discontent', this period combined very cold weather and industrial unrest. It was seen as being the last nail in the coffin in the re-election hopes of

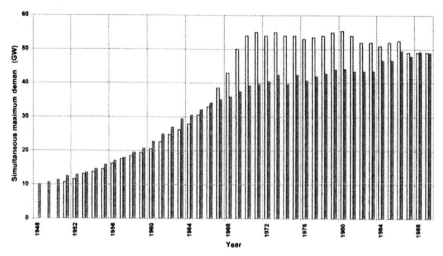

Figure 3.4. A comparison of the predicted simultaneous maximum demand (SMD) on the Central Electricity Generating Board system seven years ahead (open bars) and the actual demand recorded in each year (shaded bars).

the Labour Government, led by Jim Callaghan, and provided a platform for Mrs Thatcher's victory in the General Election of May 1979 and her policies of both confronting the unions and rolling back the frontiers of government.

The meteorological story is quite clear. After a relatively mild December the cold weather set in on New Year's Eve and with only a few breaks continued into late March. It was followed by a cold spring which did little to lift the electorate's spirits. The mean temperature of January and February was about 3.5 °C below the long-term average. Only 1963 and 1947 featured lower temperatures for these two months in this century. Furthermore, the variable nature of the weather meant there was frequent snowfall and relatively short periods of widespread severe frosts interspersed by brief thaws, which proved particularly disruptive.

On its own this combination of weather, although disruptive, was probably not sufficient to cause major economic damage. Energy consumption in the first quarter of 1979 rose over 9 per cent above the level of 1978. The electricity supply industry was sorely pressed to meet demand in spite of high capacity levels, and the transport system was seriously disrupted. The fact that the rest of northern Europe was experiencing similar cold conditions with the same surge in energy demand, while the USA was suffering from an even more extreme winter (see Section 3.5) could have made things even more difficult. At a time of growing concern about international energy supplies, the cold weather might have been expected to have precipitated a demand crisis. But because stock levels were adequate this did not occur. It was only after the revolution in Iran had run its course and war had broken out between Iran and Iraq that the next oil shock hit the western world (the average price of imported oil rose from around $13 per barrel in early 1979 to around $33 per barrel in the second half of 1980).

In the UK, the Government's problems had been building in the autumn, following the rejection by the Trades Union Congress of the proposed 5 per cent pay limit. After a series of settlements in the private sector which were well above the Government's guidelines of 5 per cent, widespread industrial action broke out early in January. On 22 January there was a 24-hour strike of 1.5 million public service workers from hospital porters to grave-diggers. This was followed by regional stoppages. The campaign of disruption resulted in the closure of hospital wards; delays in operations with union activists sometimes deciding on priorities; corpses awaiting burial in deserted warehouses; and rats scurrying through snow-covered heaps of refuse in the streets. While the part these events played in Mrs Thatcher's victory is a matter of political debate, there is no doubt that the severe winter weather reinforced the impact of the disruption in the minds of the electorate.

Since 1979 the UK has not experienced a sustained cold winter. But shorter cold spells in December 1981, January 1982, January 1985, February 1986 and January 1987 all provided timely reminders of the vulnerability of the country to bouts of snow and ice. Indeed, the less remarkable short burst of arctic weather in February 1991 produced perhaps the most memorable phrase to explain breakdowns in public services, when British Rail excused their performance on the grounds that they had been caught out by 'the wrong type of snow'. The tribulations of the water industry in Scotland and Northern England after the brief intense cold spell at the end of December 1995 shows nothing has changed.

The other side of the coin is the benefits of mild winters. In the exceptionally benign winter of 1989–90, the third warmest in the CET record, the consumption of gas (30 per cent of all UK energy consumption) was 7 per cent below trend – a major saving. So, while the winter was stormy (see Section 4.2), the reduction in domestic heating bills was a major saving for those who had not had their chimney-stacks blown down.

3.5 United States winters

As noted at the beginning of this chapter, the USA is much better equipped to handle sudden bouts of severe winter weather. But where these extremes come unexpectedly, or last much longer than usual, the system can buckle. Having lived in Washington DC for three years, I can verify that unexpected snowfall can cause chaos, especially if a snow emergency is declared and the Federal Government sends all its employees home at the same time. So the capacity of the USA to handle winter weather is a matter of degree, and the economic and political implications of cold winters can be as striking as the European examples discussed so far.

In considering the US examples, it is best to concentrate on cases where the disruption lasted for more than a few days. This is not meant to underestimate the impact of major winter storms which can cause huge damage. For example the 'Blizzard of the Century' that hit the East Coast in March 1993 caused over $3 billion damage, and the comparable storm in January 1996 may have run it close in overall costs. Indeed, the most damaging winter storm was the 'Great Appalachian Storm' of late November 1950, whose costs are reckoned to be equivalent to nearly $7 billion in current prices. But, in exploring the wider implications of cold winters, it is weeks of disruption, rather than days, which expose the fault-lines in the economic and political structures.

The eastern USA was hit by three consecutive exceptionally severe winters at the end of the 1970s. All of these caused substantial disruption. It was, however, the first one, which reached is apogee at the end of January 1977, that had the greatest impact. It came after five mild or very mild winters and was preceded by a very cold autumn, which was widely regarded as a sign of what would follow. At the same time there had been a series of warnings from experts about approaching shortages in natural gas. To make matters worse, there was a change in administration with the Democrat President, Jimmy Carter, taking over from the Republican Gerald Ford. This transfer of power took place on 20 January, smack in the middle of the worst of the freeze, and inevitably the incoming Administration was accused of being ill-prepared to handle the crisis.

The meteorological situation was a classic case of a Pacific block (see Section 5.7) with high pressure off the coast of Oregon diverting the normal westerly wind patterns up towards Alaska and then down from northern Canada into the eastern half of the USA. This combination

Figure 3.5. The massive ice build-up at Niagara Falls in early February 1977 caused by the intense cold and prolonged snowfall. (Reproduced by permission of Popperfoto.)

produced exceptionally mild weather in Alaska, a disastrous drought in California and bitter cold east of the Rockies. Having become a recurrent feature of the weather during November and December 1976, it locked in position throughout January 1977 to break many records. Overall it was probably the coldest month experienced in the eastern half of the USA in the last 200 years, but not in the country as a whole because of the relative warmth in the west. This record was claimed by January 1979 which was the coldest throughout the country, although the eastern half of the country could not match the extremes of 1977. Virtually all the USA east of the Mississippi had a monthly temperature anomaly of at least 5 °C below normal and in the upper Ohio Valley it reached 10 °C below normal (Fig. 3.5).

Because the intense cold hit the most populous areas of the country, it led to record fuel demand. When added to the problems of supplying natural gas, the disruption was inevitable. To make matters worse, the upper Mississippi and its tributaries froze solid, and barges transporting both heating oil and salt to clear the roads were marooned. So not only were alternative supplies of energy cut off but also the scope to clear the icy roads was reduced.

Against this backdrop it was hardly surprising that President Carter produced emergency legislation to enable him to have rationing power over natural gas supplies for three months. As a stopgap it was grudgingly passed by the House of Representatives and the Senate. This temporary measure did not really address the political conundrum of pricing inter-state gas supplies, which lay behind the shortages. Furthermore, before the legislation could be fully tested the weather relented and the temperatures in February rose to above normal after an intensely cold first week. The reprieve also allowed the American public to avoid the unpalatable fact that their massive energy consumption and their overheated houses lay at the core of the issue. Nevertheless, the winter exposed the vulnerability of the system, with two million workers being laid off temporarily and estimated economic losses running in the region of $20–30 billion in 1977 prices. Politically, it ensured that the Carter Administration started with 'two strikes against it'.

If many Americans had assumed that January 1977 was an aberration, unlikely to be repeated for many years, the winter of 1977–78 came as a nasty shock. While it did not feature a month quite as extraordinary as January 1977, overall the winter (December to February) was only marginally less cold. What was worse, it was punctuated by a series of major snowstorms which paralysed the major cities of the East Coast from time to time: eighteen severe snowstorms brought the greatest accumulation of snow across Illinois, Ohio and western Pennsylvania since these areas were settled at the end of the eighteenth century. Although this weather caused massive disruption and significant loss of life, it never reached the level of its immediate predecessor. Nevertheless, in Illinois alone the estimated cost of damage ranged as high as $2 billion.[9]

In meteorological terms the winter of 1978–79 was even more exceptional. In December the cold hit the west of the country, where in many places it was the coldest on record. The cold moved eastwards so it covered the whole country in January, which was the coldest on record for the entire country. By the end of the winter it had migrated to the East Coast where many stations notched up the second or third coldest February on record. But, in spite of it being by far the coldest winter across the US in the last 100 years, it did not hit the most populous regions as hard as the 1976–77 winter. Even so there was evidence of declining impact, or increasing adaptability of society. Even when the Midwest was hit by the 'worst blizzard in memory' in mid-January killing at least 100 people, and Chicago was effectively shut down for a week, there was a sense that such weather was no longer unusual and that increasingly these extremes were something that could be taken in society's stride. So, although the eastern half of the USA had experienced a series of winters which might, on the basis of chance, be expected to occur once every five hundred to a thousand years, there was a sense that it was becoming easier to handle with increasing experience.

A more lasting consequence of this trio of bitter winters was to reinforce a general outward migration from the north-eastern USA. Usually defined as the drift from the Rust Belt to the Sun Belt, this movement was initially driven by the decline of the traditional heavy industries of the north-east. On top of other economic and social factors the crippling winters of the late 1970s proved to be the last straw for many. It will be interesting to see whether these meteorological factors are reversed if those who moved to the coast of the Gulf of Mexico suffer many more hurricane seasons like that of 1995.

Since the 1970s US winters have been less extreme, although there were a number of notable sustained cold spells. Even so it is estimated that the economic consequences of the two exceptional cold waves that hit the eastern half of the country during January 1982 caused several billion dollars worth of damage. Similarly, the dramatic cold spell of December 1983 broke many records and caused widespread disruption and some $2 billion damage to the Florida citrus industry.[10] Nevertheless, there is a sense that the frequent bouts of cold weather did ensure that systems put in place to handle such extremes were working more effectively. Further evidence of the economic consequences of this increasing adaptability of US society came in December 1989. A study of the economic impact in the Lake Erie snowbelt, where it was the coldest December in the last 100 years, showed that there had been remarkably few losses.[11]

Recent winters in the USA have been a mixed bag. The winter of 1991–92 was the warmest in the last 100 years across the country, while that of 1994–95 was the third warmest. In between, 1993–94 was one of the five coldest winters this century in the Great Lakes region. The two cold waves in January 1994 broke many all-time low temperature records and provided a chilling reminder of just how bitter US winters can be. Whether or not the relative warmth in recent years is a belated reaction to global warming or merely a lull before a few more sustained arctic blasts, the message for North America is clear – do not drop your guard. Having learnt from bitter experience how to live with extreme winter weather, people would think it perverse if they had to start all over again when it next comes knocking on their door.

Talking of perverse reactions to cold winters, the response of Wall Street to the consequences of the huge snowstorm in January 1996 takes a lot of beating. This blizzard, together with other snowfall, paralysed much of the north-east, and temporarily threw many people out of work. So the unemployment figures rose in January, and fell in the next month. When the February figures were announced in early March, they showed a rise of over 700 000 in the number of people in work. This ostensibly good news was greeted with horror on Wall Street and the Dow Jones Index fell 171 points (3 per cent) in a day. The reasoning was that the economy might be 'overheating' and interest rates and inflation might be about to rise. For the traders playing arcane games on the bond markets and the dollar/yen exchange rates, this meteorologically induced blip produced sudden, albeit temporary, panic. All of which shows you should never underestimate markets' ability to put a different spin on uncertainties resulting from aberrant weather.

3.6 Notes

1 Alanbrooke's views are recorded in diaries and autobiographical notes which are quoted in Bryant (1957), pp. 65–67.

2 Liddell Hart (1970).

3 Stolfi (1980).

4 Neumann (1992).

5 Hannah (1979).

6 Kaiser (1974).

7 Central Electricity Generating Board (1963), p. 15.

8 Ibid., p. 57.

9 Oliver (1981) pp. 60–65.

10 *Billion Dollar U.S. Weather Disasters 1980–1996*. National Climatic Data Center, Ashville, NC, report 19 January 1996.

11 Schmidlin (1993).

4

Storms, floods and droughts

'The time has come', the Walrus said,
'To talk of many things:
Of shoes – and ships – and sealing wax
Of cabbages – and kings –
And why the sea is boiling hot –
And whether pigs have wings.'
Through the Looking-Glass, Chapter 4

It may seem a little perverse to pay so much attention to cold winters, given global warming is in the forefront of people's thinking. It is worth recalling, however, that in the late 1960s and early 1970s the possibility of global cooling was in the forefront of many climatologists' thinking. Subsequent climatic events have completely reversed this perspective and now the vast majority are now concerned with warming rather than the next Ice Age. The scale of this shift in opinion and the messages about the non-linear response of economic and social systems to extreme weather make the examples of cold winters of relevance, whether or not they will become a less frequent part of our future.

When it comes to current climatic developments most people are worried about the mixture of storms, floods and droughts that seem to be a growing feature of the weather. Central to the debate is the whole question of whether global warming is producing a more extreme weather and what the role of the oceans will be in these changes. Of particular interest are the quasi-periodic fluctuations in the equatorial Pacific Ocean (the El Niño Southern Oscillation, ENSO), the associated changes in sea-surface temperature (SST) and rainfall patterns throughout the tropics, and their possible connections with weather patterns at higher latitudes. At the same time,

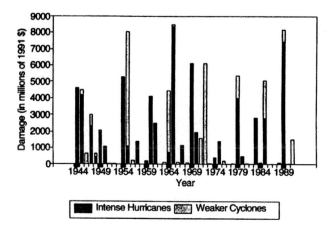

Figure 4.1. Tropical cyclone damage in the United States (in millions of 1991 dollars) from tropical cyclones. (From Landsea, 1993.)

the growing awareness that variations in the way the oceans transport energy from the tropics to polar regions (see the Great Ocean Conveyor Belt, Section 5.8) has added a new dimension to what could be driving current the changes in the global climate. But, while the climatic plot thickens, what really concerns the general public is the errant weather.

4.1 A crescendo of hurricanes?

To many people the exceptionally busy hurricane season in the tropical Atlantic and Caribbean in 1995, and the above average incidence of tropical storms in 1996 were further confirmation of the impact of global warming. The 19 named tropical storms and 10 hurricanes in 1995 was the second highest figure in records going back to the late nineteenth century. Since Gilbert (the most intense storm on record in the region) rampaged through the Caribbean in 1988, first Hugo in 1989 and then Andrew in 1992 brought huge damage to the USA. The $20–30 billion costs of Andrew were seen as the latest manifestation of a rapidly escalating trend in insurance costs (Fig. 4.1).

The simple argument for global warming leading to an increase in the number and intensity of hurricanes is that they are fuelled by the heat and moisture available in the tropical oceans. Broadly speaking, one essential ingredient for a developing hurricane is SSTs above 27 °C. So if the oceans get warmer, in theory there will be more energy to produce and sustain more and bigger hurricanes. Given the global warming trend this century (see Fig. 5.7), on the face of it there should be an increase in hurricane activity. More detailed analysis provides little support for this conjecture. The current generation of computer models (see Section 6.2) are not capable of providing reliable estimates of how the incidence of tropical cyclones will be altered by global warming. Various studies have produced equivocal results and the conclusion of the Intergovernmental Panel on Climatic Change (IPCC) is that the issue of whether global warming will produce an increase or decrease in these storms is very much an open question.[1]

The statistics tell a different story. Although coverage was less complete in the first half of this century, it is clear there is no trend in the number or intensity of hurricanes in the tropical Atlantic (see Fig. 1.1). Since 1944 the US Air Force and Navy have continually flown missions to monitor hurricanes, and from the mid-1960s weather satellites have provided an even more complete picture. These observations clearly show there has been a marked decline in activity in general, and in the incidence of intense hurricanes in particular during the last 50 years. From year to year there have been sudden switches from intense activity to quiescence. The most notable feature is that hurricanes were more frequent between the mid-1940s and the end of the 1960s. So for the time being, it is probably wise to assume that the upsurge in 1995 and 1996 is little more than a fluctuation in what is a highly variable phenomenon.

Meteorologists seeking to explain the incidence of tropical cyclones have concentrated on how the atmosphere is influenced by the pattern of sea surface temperatures throughout the tropics. In addition, the influence of prevailing wind patterns in both the troposphere and stratosphere, together with humidity at different levels, are seen as important factors. Links between the ENSO, rainfall in the Sahel region (see Section 4.5), the quasi-biennial oscillation (QBO) in the stratosphere (see Section 5.7) and the temperature of the tropical Atlantic are central to improving our understanding. William Gray and colleagues at Colorado State University have been using these and other changes to predict hurricane activity in the Atlantic for a number of years, with considerable success (see Section 7.5).[2]

What all this shows is that the links between global warming and hurricane activity are not simple and hence predictions about their future

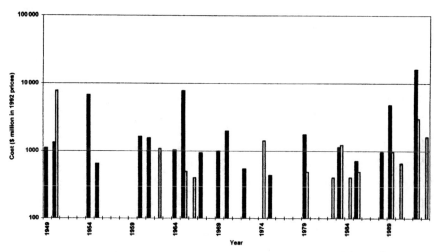

Figure 4.2. Insured costs of major weather disasters in the United States for both hurricanes (solid bars) and also other storms, tornadoes and floods (open bars). (Data from Nutter, 1994.)

economic consequences must be treated with caution. Trends in the economic costs of hurricanes in the USA, which have dominated thinking on the possible consequences of global warming, need to be handled with even more care. Although there can be no dispute that the costs to the insurance industry are real, what they measure and how their variation over time can be compared is much more difficult. The standard practice is to reduce damage figures to a common base by correcting for the changing cost in construction and the changing population. For instance, between 1950 and 1990 construction costs in Miami rose by a factor of 5.7 and the population rose by a factor of six. So to convert the cost of a 1950 hurricane with a comparable event in 1990 we need to normalise the earlier figure by the product of the rise in construction costs and the rise in population (i.e. $5.7 \times 6 = 34.2$). This results in, say, Hurricane King, which caused $28 million of damage in 1950, being estimated as equivalent to $957 million in 1990 prices.[3] The product of this type of analysis for Atlantic hurricanes striking the USA is shown in Fig. 4.1. Although the analysis does not include the huge losses caused by Hurricane Andrew, it is still reasonable to conclude that the real level of damage has not risen markedly in the last 50 years.

An alternative set of figures provided by the US insurance industry for major weather disasters, including both hurricanes and other storms, provides a slightly different insight (Fig. 4.2).[4] Because flood damage is often not covered by insurance, some hurricanes show up less strikingly (e.g. Diane in 1955, Camille in 1969 and Agnes in 1972). Nevertheless, even with the inclusion of Hurricane Andrew, there is no marked trend, but an increasing vulnerability to single huge disasters. While hurricanes represent around 70 per cent of the losses caused by major weather disasters in the USA, the figures for other events (see Fig. 4.2) provide further confirmation of the spotty nature of losses and lack of the clear trend. The fact that by far the most costly disaster was the Great Appalachian Storm in November 1950, which resulted in insured losses of $174 million (in 1992 figures this is estimated to be equivalent to $6.6 billion), shows the USA has a long history of damaging weather. Moreover, efforts to compare losses over the years may underestimate the rate at which insurance cover has risen to reflect the increasing affluence of people who choose to live in the coastal areas of the south-eastern states of the USA. In this context the regulation of the growth of vulnerable shore-line developments is an important factor. The US insurance industry reckon, that proper building codes and their enforcement could have reduced insured losses in Hurricane Andrew by 30 per cent.[4]

The observations show that analysing trends in weather-damage insurance costs and the statistics of extreme events is no easy matter. Since, by definition, extremes only happen rarely, establishing changes in their frequency is bound to take time. This means while there is no doubt about their individual impact, where they stand in the longer term scheme of things is more perplexing. In the case of hurricanes in the USA, the exceptional feature of Hugo, and much more so in the case of Andrew was the fact that they both hit populous areas. Given that the path of maximum damage is, at most, a few tens of kilometres wide, it is a lottery as to how populous an area a hurricane will strike. The statistics of hurricane activity and the economic damage caused have to be scrutinised very closely before drawing any conclusions about what causes the changes and what they mean for the future.

4.2 Mid-latitude storms

More than any other weather event in recent decades, the 'Great Storm' which hit south-east England on the night of 15–16 October 1987 was instrumental in arousing British public and political interest in global warming. The reason it is so embedded in the national psyche is the product of an interesting combination of factors. The bald facts are that

Figure 4.3. Damage caused by the Great Storm of October 1987. (Reproduced by permission of Surrey Herald Newspapers.)

because it came in the early hours of the morning the death toll was relatively low, with 20 killed, but the economic damage based on insurance claims exceeded £1.2 billion ($1.9 billion) (Fig 4.3). More memorable to many people was the damage to woodlands with over 15 million trees blown down in south-east England. Many of these trees were well over a hundred years old and seemed a permanent feature of the landscape, which made their destruction so much more distressing. Furthermore, it was immediately followed by the worldwide stock market crash, which reinforced its apocalyptic nature in many people's minds. Then there was the failure of the Meteorological Office to predict the intensity of the storm, which was compounded by the comments of Michael Fish, a well-known TV weather presenter. At the 1.25 p.m. weather slot on the BBC on the day before the storm, he said: 'Earlier today a woman rang the BBC and said she had heard there was a hurricane on the way. Well, if you are watching, don't worry, there isn't.' This wonderful example of famous last words has become part of British weather folklore. Furthermore, despite attempts of meteorologists to explain that it was not truly a hurricane, to the general public it will always be known as 'the hurricane'.

The failure to forecast such a cataclysmic event led to a public outcry. Although it was recognised that more accurate warnings would have made little difference to the scale of damage, the state of unpreparedness was laid at the door of the Meteorological Office. The UK Government set up a formal enquiry into what went wrong. The report by Professor Robert Pearce and Sir Peter Swinnerton-Dyer[5] concluded there was no question of negligence on the part of the Meteorological Office, nor were its forecasts worse than other agencies, in spite of the fact that on this occasion the French model did a better job of predicting the path of the storm. Perhaps the most important conclusion was that the lack of observations in the vicinity of the storm when it was over the Bay of Biscay was a major factor in the poor performance of the final forecasts. Knowledge of the initial state of the atmosphere is central to all weather predictions and will be considered further in Chapters 5 and 6.

When on 25 January 1990 another intense depression ran across the southern half of the country killing 47 people and causing nearly £2 billion ($3.2 billion) in damage to property, it was widely seen as confirmation that storms of this type were becoming a more common feature of the climate because of global warming. This storm, sometimes known as the Burn's Day Storm, featured comparable wind strengths to the Great Storm of October 1987, but over a much wider area.[6] The damage to trees was, however, much less as deciduous species had no foliage on them. Throughout the rest of north-western Europe the storm killed over 50 people and did a comparable amount of damage to that incurred in the UK. As for the forecasts, the Meteorological Office redeemed itself by both accurately predicting the development of the storm at least four days in advance and then issuing copious warnings as it approached. The sense of increasing storminess was reinforced by the damaging severe gales that were a feature of the following month.

As with hurricanes in the USA, the same problems arise of interpreting the statistics of extreme events. Again the long-term statistics do not show any clear trend. Mike Hulme and Phil Jones of the Climatic Research Unit at the University of East Anglia have analysed sea-level pressure patterns since 1881.[7] Their results are shown in Fig. 1.2. A similar analysis of atmospheric pressure maps to estimate seasonal windspeeds since 1876 by the German Weather Service in Hamburg[8] has reached the same conclusion. Using statistics from German Bight in the southern part of the North Sea, they showed no evidence of trends in windspeeds in any season. The one relevant statistic that does show a marked increase is the number of winter Atlantic depressions with central pressures below 950 mb.[9] These major storms are an indication of the strong westerly circulation in the late 1980s and early 1990s. Their economic impact is, however, less easy to gauge given that many of them lived out their existence far from land.

A more important factor in assessing recent trends in storminess is that it is possible to point to many other severe storms that have caused great damage across the British Isles and northern Europe. They range from the famous storm recorded by Defoe in 1703 to that of January 1976. Most notable of all is the intense depression which produced the storm surge on 31 January 1953 that flooded the east coast of Britain and breached the Dutch dykes. This caused far greater loss of life than recent storms and did immense damage to Scottish forests. While London was spared the worst of this flood, the fear of a devastating combination of high tides and an exceptional storm surge led to the building of the Thames barrage. This was eventually opened in 1983 and was almost immediately pressed into service when precisely 30 years after the disaster of 1953 a similar storm situation developed. In the Netherlands, the loss of over 1800 lives and the massive damage to farmland and property in 1953, led to decades of strengthening of the dykes and the building of major flood control systems in the Rhine and Scheldt estuaries.

All of this shows that in north-west Europe there is a long history of grappling with the threat of winter storms that sweep in from the North Atlantic. Moreover, where the costs are unacceptably high, governments have taken on the burden of building defences to ward off catastrophe. In terms of the challenges which have to be confronted, the storms of October 1987 and January 1990 are the latest chapters in the saga. As yet there is no evidence to suggest they are part of a significant trend, or that their occurrence can be attributed to global warming. Indeed, thus far their real significance may be political, in that they influenced Mrs Thatcher's Conservative Government to take the threat of global warming more seriously. This is a good example of where the historical impact of events may be all a matter of how they are perceived rather than what they really represent.

4.3 Floods

Since the beginning of recorded history floods have played a special role in the development of many societies. Whatever the origin of the legends of the Biblical Flood or Gilgamesh, there is no doubt that the benefits of exploiting the fertility of the floodplains of the world's great rivers has always had to be balanced against the risks of loss of life and property when major floods struck. Events in recent years have shown that this age-old truth remains as certain as ever. But what matters is whether the pace of events and their cost is changing significantly. If so, is this a consequence of shifts in the climate or more to do with the management of river flows and exploitation of floodplains around the world?

The arguments about the causes of major floods and their economic impact can be covered in considering three examples of recent inundations: the summer of 1993 on the upper Mississippi, the winter floods of December 1993 and January 1995 on the Rhine, and the perennial problems of Bangladesh. Starting with the Mississippi floods, there is no doubt that the meteorological conditions were exceptional.[10] After a wet autumn and a snowy winter which left Iowa with its greatest snowpack since the spring of 1979 (see Section 3.5), the rain started in earnest in April. For the Mississippi watershed the 1993 precipitation was the greatest since records began in 1895 for not only the April to July period, but also for May to July, June and July, and July on its own. Many places in Iowa and Kansas had more rain in these four months than in a normal year, while from North Dakota to Illinois the totals for June and July broke all-time records, often by wide margins. The chance of these extremes recurring was estimated to be one in 200 to 1000 years.

Inevitably the flood levels broke all records. Severe flooding began in May on the Redwood River in Minnesota and in June on the Black River in Wisconsin. This was followed by record levels on the Kaman, Mississippi and Missouri Rivers in July. At St Louis the water crested 1.9 metres above the previous record set in 1973 and exceeded the earlier figure for over three weeks. Just north of the city the floodwaters reached a width of 32 km where the Missouri joins the Mississippi (Fig. 4.4). Over 7 million hectares (nearly 20 million acres) was flooded across nine states, and at least an equal area was saturated, which further added to crop losses. At least 50 000 homes were damaged or destroyed and 85 000 residents had to evacuate their homes. In Des Moines, Iowa, the residents were without potable water for 12 days. Some flooding was caused by levees collapsing under the sustained pressure of water whereas in other places it flowed over the top. Overall some 58 per cent of the 1400 levees on the Mississippi and Missouri Rivers were overrun or breached by the water.

The immediate economic consequences were estimated to amount to

Figure 4.4. Satellite image of Mississippi floods of July 1993. (Reproduced
by permission of Radar Satellite International.)

some $15 billion. The death toll was mercifully small, totalling 48. Over
4 million hectares (10 million acres) of farmland was flooded and crop
losses exceeded $5 billion. Many farm animals perished. The crop losses
can be inferred from the fact that the national soybean yield was 17 per
cent below the record crop level of 1992 while the corn (maize) yield
dropped 33 per cent. The damage to housing, property and business made
up the remainder of the estimated costs.

The longer term impact of the floods is more instructive. The scale of
the damage led, for the first time, to serious questions about the strategy
of flood control. The assumption that the best response to successive
meteorological extremes is to build bigger and better defences was called
into question. It was argued that it would be better to accept the inevita-
bility of occasional floods and plan activities around this assumption. Not
only would this prove more economic but it would have advantages in
managing heavy rainfall; controlled inundation of the floodplain could lead
to better watershed management and a slower run-off, which would
reduce peak flows and hence flood levels. These issues are still the subject
of intense debate and involve difficult political decisions about individuals'

freedom to live in vulnerable situations and the obligations of the state to provide them with adequate protection.

The scale of federal disaster aid led, however, to prompt action. The Federal Government decided to finance a scheme to 'retire' the most vulnerable riverside properties. Local towns were funded to purchase and demolish the most frequently flooded properties and turn the areas into parks or recreational land. In the state of Missouri alone $100 million was used to purchase 2000 residential properties. It was reckoned that this measure alone would save $200 million over 20 years even without exceptional flooding. At the same time insurance schemes were modified to require that people with federally backed mortgages were adequately covered. Furthermore, the ability to take out insurance to provide cover within five days, when flooding was imminent, was blocked. When near-record floods struck some of these areas only two years later in 1995, these measures looked like good value for money. Whether decisions of this type become widespread depends on having reliable evidence as to whether such extremes are becoming more common and, if so, why (see Section 7.7).

The same issues arise in the case of the Rhine floods. In both December 1993 and January 1995 much of the region drained by the Rhine and its tributaries experienced exceptionally heavy rain. During the first wet spell, parts of the region had three times the average December rainfall. Towns like Cologne and Koblenz suffered their worst floods with water levels coming within 6 cm of the level reached in the great flood of 1926, which was the highest in the last two centuries.[11] Belgium, eastern France and the south-east Netherlands were equally badly hit. The costs in Germany alone were estimated to run to $580 million.

Just 13 months later a similar situation occurred. The Belgian Ardennes and parts of northern France had the wettest winter this century. In Cologne the flood level equalled the record of 1926. Downstream the peak flows were less extreme, but were sustained for longer. So the Netherlands suffered the worst conditions. For a while it was touch and go as to whether the floods would undermine the dykes holding back the North Sea and lead to a much greater disaster. For safety, 200 000 people were evacuated from their homes. Fortunately, the dykes held, but two such severe floods so close together raised a lot of questions about whether this was part of the warming trend in the climate. They also demonstrated how people adapted to the threat. Although the flood levels were higher in Germany in January 1995 the economic costs were halved because

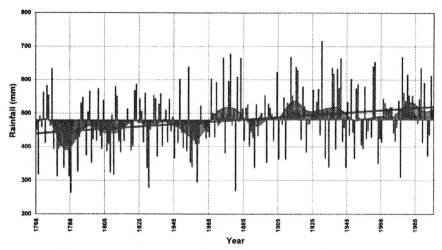

Figure 4.5. Rainfall figures for England and Wales for the winter half of the year (October to March) between 1765 and 1995, showing a steady increase over the period, together with smoothed data showing longer term fluctuations. (Data from Wigley *et al.* 1984, plus updating from statistics published regularly by the UK Meteorological Office.)

people heeded the warnings and protected their property more effectively.

The interpretation of these events has to be in terms of the combination of meteorological trends and greater exploitation of major rivers which lead to a tendency to make such disasters slightly more likely. In north-west Europe there has been clear upward trend in rainfall in the winter half of the year, while there has been an almost equal and opposite downward trend in summer rainfall (Fig. 4.5). These changes have to be set against other social developments which tend to amplify the economic impact of extreme weather. So while the rainfall that produced both floods was unusually heavy, the changes in the drainage in the Rhine watershed and in channelling the river to make it easier to handle barge traffic played as big a part in the disasters. The speeding up of the run-off has approximately halved the time that heavy rainfall will take to flow through the system. On the positive side, people learn from experience, and forecasts improve, so they are able to take better action to reduce losses.

The same issues emerge when considering floods in the developing world. Nowhere are these more evident than in Bangladesh. Here flooding is endemic, with 80 per cent of the country in the flood plain of the Brahmaputra, Ganges and Meghina rivers with no point higher than 30 m.

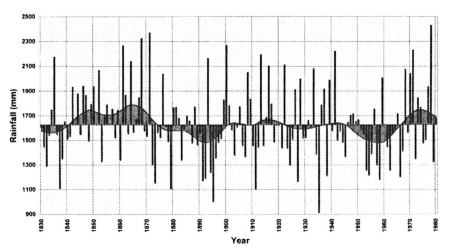

Figure 4.6. Annual rainfall for Calcutta from 1830 to 1980, showing no significant increase in recent decades together with smoothed data showing longer term fluctuations.

In heavy monsoon years such as 1987, 1988 and 1993, virtually all of this land was flooded. But in almost any year with above average rainfall widespread flooding can be expected. Furthermore, 40 per cent of the land is only a metre or less above mean sea level, and hence is acutely vulnerable to storm surges associated with tropical cyclones moving in from the Bay of Bengal. In almost every aspect of global warming Bangladesh is a test case of the consequences of climatic change and the benefits of interventions to reduce flooding.

In terms of climatic change there is little evidence of a marked trend in rainfall in the Ganges delta (Fig. 4.6) since the early nineteenth century. There has, however, been a marked increase in flooding. This appears to be more to do with accelerated drainage in the foothills of the Himalayas. The rate of clearance of the forests of Nepal, Bhutan and Assam for both cultivation and firewood in recent decades is claimed to be a major factor not only in causing more rapid run-off but also greatly increased top-soil erosion. Both these changes have contributed to the increase in inundations. But, the absence of reliable long-term statistics makes it difficult to identify which particular aspect of the changes upstream is most important and how best to tackle it.

In the case of tropical storms, the disaster of 1970, when some 300 000 people perished in a tidal wave that swept across the deltas of the Ganges and the Meghina, galvanised national authorities into action. An innovative preparedness programme was developed using local volunteers to disseminate cyclone warnings and guide the local populace to constructed mounds or cyclone shelters. During the 1980s the low level of casualties was seen as evidence that this programme was working.[12] The loss of life in the cyclone of April 1991, when some 130 000 people died, provided a stark reminder of the vulnerability of Bangladesh. While there is no clear trend in the incidence of cyclones in the Bay of Bengal, the combination of population pressures to exploit the rich soil deposited in the delta regions, and rising sea levels, means that the risk of major loss of life continues to rise.

Faced with this catalogue of disasters, Bangladesh must be regarded as a prime candidate for international action. Recent events show, however, just how difficult it is to reach agreement on the best way to tackle the nexus of problems facing the country. Central to this debate is the question of funding the huge Flood Action Plan by the World Bank. Following the floods of 1987 and 1988 the Bangladesh government wanted to build a huge set of embankments to tame the Ganges and Brahmaputra rivers, protecting cities and increasing crop production. Costing $10 to 15 billion, this scheme has run into opposition from local political groups and western environmentalists and hydrologists. Opponents argued that not only would the scheme make matters worse for the vulnerable communities in the delta regions, but also it would isolate the rivers from their flood plains and hence greatly increase the damage if the embankments were ever breached. The experience of the Mississippi in 1993 shows that even in the most advanced countries taming mighty rivers is a vain hope. The plan would also cut off extensive wetlands, damaging fisheries which are the principal source of animal protein in Bangladesh.

The associated issue of soil erosion is an interesting example of how complicated life is. Leaving aside the damage it does to the local environment, it has a variety of impacts downriver. First, in silting up waterways, the sediment plays a significant part in causing floods. In the delta the huge quantities of material swept downriver (1.5 to 2.5 billion tonnes per year) creates new land. The net balance is a complicated mix between deposition, settlement and coastal erosion. So it is not possible to estimate either future changes in land area or how they will be affected by sea-level rises. But, while this process may benefit those living in the delta areas, overall the effect of soil erosion is immensely damaging to the Himalayan region. More generally, it has been calculated that soil erosion by wind and water cost the USA $44 billion per year, while the global annual costs are $400 billion.[13]

These figures show the scale of what is at stake. Altering the management of the hydrology of the catchment areas in the Himalayas, with its questions of deforestation and soil erosion, can only be solved through international negotiations. Finding an acceptable balance of interests for the countries involved will not be easy. This example does, however, underline the essential nature of confronting climatic change and other environmental issues as a whole. Only by identifying the total impact of current practices, the true extent of common interests and the potential mutual benefits of concerted action will it be possible to negotiate reasonable compromises between countries.

4.4 Droughts

All around the world droughts are the obverse to floods on the climatic coin. Just as exploitation of the floodplains of many rivers carries the risk of inundation, so the farming of more arid places includes the ever-present threat of drought. Moreover in many parts of the world the comfortable zone between these two extremes is narrow, so that farmers may have to survive adjacent periods of drought and flood with little respite in between. This is true of many parts of northern Europe and North America. Indeed long before this climatic truth became readily apparent to European farmers, the Egyptians recorded the various levels of the Nile floods: the greatest floods were defined as 'disaster', more modest inundations declined from 'abundance' and 'security' to 'happiness', and dry years brought 'suffering' and 'hunger'.[14]

So drought has always been part of the human condition. As with storms and floods, a few recent examples will serve to illustrate the economic impact of drought in the context of current concern about climatic change. They also serve to bring out how certain aspects of the fluctuations from year to year in the weather in some parts of the world are more predictable than in other places. In particular, this is the area where the connection between tropical sea surface temperatures and rainfall patterns has become well established.

But, before turning to the tropics, an interesting starting point is the events of 1972, and, in particular, the harvest failure in the Soviet Union. Not only did this year show up the vulnerability of the world's grain supplies, but it also awakened scientific interest to the importance of the tropical Pacific to world-wide weather patterns. The combination of the collapse of the anchovy harvest off the coast of Peru, the drought in Sahel, and the driest monsoon season since 1918 in India, together with events in the Soviet Union, first brought the El Niño into the limelight. Although the full significance of these global teleconnections was not to become more widely appreciated until little over a decade later, the aberrant weather of 1972 triggered new meteorological awareness into understanding what controls these fluctuations.

The drought in the grainlands of the USSR of 1972 marked an important turning point in Soviet agriculture. Drought and crop failures are not new to Russia. Eisenstein's 1929 film, *The Old and the New*, had a memorable scene in which peasants led by priests form a great procession to pray for an end to the drought that afflicts their land. The film contrasted this approach with the success of the technology of the modern Soviet state: the triumph of modern technology over ancient superstition. In truth, to the extent that Eisenstein's optimistic message proved correct, it was the result of greater and more reliable rainfall in the Soviet grainlands between the 1920s and the 1950s. But, under Stalin, after the brutal period of 'collectivisation' in the 1930s, Soviet agriculture was starved of investment. At the time of his death, 1953, food production per head of population was no higher than in 1928.

During the 1950s and 1960s the system made up for lost time, and both production and productivity rose. Much of this expansion was associated with the opening up of the 'virgin lands' of central Asia. But these are areas of low rainfall which fluctuates dramatically from year to year. The downturn of rainfall in the 1960s made matters worse. There were poor harvests in 1963 and 1965, when for the first time, the USSR had to import grain. Good harvests in the late 1960s brought a steady rise in output and a feeling that all was well. The failure in the traditional grainlands of the Ukraine and the Russian Federation in the hot, dry summer of 1972 came as an awful shock. Although the drought was only on a par with earlier bad years, the summer was the hottest this century. Ironically, however, complete disaster was staved off by the good harvest in the new lands of north Kazakhstan. The real impact of the shortfall was, however, that the Soviet Union had to go on to the world grain markets in a big way. In 1972 and 1973 they imported some 30 million tons of grain to meet needs, in spite of a record-breaking bumper harvest in 1973. The fact that their buyers managed to achieve this without initially causing a sudden and large rise in prices has, in some quarters, come to be known as the 'Great Grain Robbery'. When the scale of the purchases became apparent, prices rose during 1973 to levels that would not be seen again until 1996. This buying coup meant that henceforth the Soviet Union was recognised as a major importer of grain – never again would the Chicago Futures Market get stung the way it did in 1972.

The subsequent history of Soviet agriculture shows that the bad weather was only part of the problem. In 1975, however, the weather did ram home the message about the vulnerability of the system. Both the new lands and the traditional grainlands were hit by drought and production fell 70 million tons short of the target set by the Ninth Five-Year Plan. During the late 1970s and early 1980s the chronic weaknesses of Soviet agriculture became more and more apparent: a combination of poor harvesting, transport and storage systems meant that even in good years much of the harvest was wasted. Furthermore, the assumption that the dry years in the 1960s and early 1970s marked a more lasting shift in the climate proved premature and, if anything, spring and summer rainfall trends have moved upward this century. So, it is hard to draw any hard and fast conclusions about just how important the weather was in exposing the inefficiencies of the Soviet system given the highly variable seasonal rainfall in the new lands. Suffice it to say that after the problems of 1972 the old triumphalism of central planning never rang true, and subsequent bad years only served to reinforce this message.

Sustained below-normal rainfall in parts of the world which have adequate year-round precipitation has a more subtle impact. The prolonged period of mainly dry months in the north-east USA from late 1961 until the beginning of 1967 is a good example of this phenomenon. A whole series of measures was taken to alleviate shortages, ranging from asking restaurants in New York City not to serve water unless asked to do so, through to the banning of watering of lawns, to protracted interstate negotiations between New York and Delaware over managing the flow of the Delaware River which provided water to both states. The overall economic impact of several years of low rainfall is difficult to measure because of the erratic nature of the ups and downs in monthly figures over such a prolonged period. Furthermore, the arrival of heavy rain through much of 1967 meant that the economic effects slowly dissipated, as did any resolve to keep water conservation measures in place.

A more clear-cut example of the impact of sustained low rainfall occurred in Britain in 1976. This drought can rightly be regarded as producing a sea-change in the British attitude to water supplies and the consequences of water shortages. While much of south-east England regularly suffers from inadequate rainfall for many forms of temperate agriculture, unless supplemented by irrigation or watering, this climatological fact was usually overlooked because of the country's rainy reputation. At the same time the temperatures involved are modest compared with summers in the eastern half of North America. A hot summer in central England is on a par with normal conditions in Caribou, Maine or central Ontario, just south of the Hudson Bay. But what matters is departures from the normal and how these disrupt standard activities. Moreover, as with other aspects of extreme weather events or climatic change, the real economic consequences are associated with changes in perceptions rather than the immediate costs of the events.

The bald facts about 1976 are as follows. The three months of the summer (June, July and August) were the hottest in the Central England Temperature record – just edging out 1826 for the top spot.[15] Moreover, the 16 months to August 1976 were the driest period of this length in the England and Wales records stretching back to the early eighteenth century.[16] As such, it can be rated as at least a one-in-250-year event. The effects of the drought built up slowly. For water authorities the failure of winter rainfall to replenish the reservoirs meant that anything but a wet summer was bound to cause trouble. For farmers, the dry winter and spring were seen, however, as a boon. But as rainfall remained below average and the heatwave set in June the picture changed. The first problems were with widespread scrub fires and shortages of water. Then the problems for agriculture started to emerge. Initially, the concerns were with the shortage of grass, hay and silage for animal fodder. Then it became apparent that crop yields would be much reduced. In the event, grain yields were some 30 per cent below trend, and root crops were hit even harder. The cost to farmers and growers was estimated to be some £400 million ($640 million) in 1976 prices. The water shortage also hit industry as supplies were cut, and there was a significant reduction in overall output.

The hidden costs of the drought were, however, of much greater economic importance. These were the damage to housing. The extreme drying out of the soil exposed the fact that building practice was not adequate to handle subsidence, especially on heavy clay soils. The cost of repairing and underpinning domestic properties amounted to well over £100 million ($160 million) in 1976 prices. As a consequence, both the costs of insuring housing and the building regulations for foundations underwent lasting change, thereby imposing significant additional costs on owners of domestic property.

As for the headline-grabbing issue of the problems of failing to meet demand for water, the story is equally complicated. In spite of great argument at the time, the whole issue of metering water, or of building a national grid to move water around the country, remain unresolved. The above-normal frequency of dry summers in the UK since 1976 has ensured that these issues have remained in the public eye. The discomfort of the privatised water utilities during the drought of 1995 has only served to heighten this debate. The obvious conclusion has to be that the economics for such investment remains open to question and every time it starts to rain again the issue is quietly shelved. This procrastination is aided by the meteorological oddity that dry summers are often followed by wet autumns. So when 1976 was followed by a drenching September and October the sense of urgency drained away. A similar pattern prevailed in both 1994 and 1995. But, if hotter, drier summers become part of the British way of life, and domestic consumers continue to insist on having traditional bright green lawns, the pressure for metering will grow.

The messages emerging from both the 1960s' drought in the north-east USA and that of 1976 in Britain are an odd combination of short-term panic measures and long-term complacency. Although there were permanent changes in certain aspects of how the risk of drought was managed, for the most part the general approach was to assume that the drought was a temporary aberration and things had now returned to normal. Only where a specific group can transfer the risk to a wider group of consumers (e.g. through the insurance market, or through changing regulations) does a more permanent change occur. Even then, where suppliers can compete for customers, a few wet years may well lead to, say, more competitive rates of insurance. So the ability to predict whether different forms of extreme weather are becoming more or less frequent is central to establishing a sensible approach to the economic challenges of climate change.

The droughts and heatwaves in the Midwest USA in 1980 and 1988 provide a slightly different perspective on the political impact of drought. These summers attracted a great deal of comment in the context of predictions of global warming. But, as noted in Chapter 2, they did not constitute climatic stress of the same order as the Dust Bowl years of 1934 and 1936. Nevertheless, the costs of the hot weather in 1980 were estimated to have run to some $20 billion in 1980 prices[17] and the costs in 1988 may have been even greater given the severity of the heat waves in urban areas of the eastern USA. So this is another reason for finding out whether such events are liable to become more frequent, although in recent years it has been excessive spring and summer rainfall which has had the greatest impact on the Great Plains (see Section 4.3).

4.5 Global connections

Fluctuations in summer weather in the USA do, however, provide a good point for stepping into the question of the links between weather extremes in mid-latitudes and interannual fluctuation in the tropics, notably the equatorial Pacific. These links centre on the ENSO, which is clearly the most significant feature in the variability of tropical weather. It constitutes a widescale ocean–atmosphere interaction which exhibits quasi-periodic behaviour. The atmospheric component was first analysed by Sir Gilbert Walker in the 1920s and 1930s. He observed that 'when pressure is high in the Pacific Ocean it tends to be low in the Indian Ocean from Africa to Australia', and termed this behaviour the Southern Oscillation.[18] Subsequent meteorological research[19] has confirmed the global scale of this phenomenon and has shown that it is associated with substantial fluctuations in rainfall, tradewind patterns and sea surface temperatures (SSTs) throughout the tropics.

The associated changes in SSTs had been common knowledge in Peru for centuries. One feature of the phenomenon of El Niño is the warm current that flows southwards along the coasts of Ecuador and Peru in January, February and March. Its onset, which brings the local fishing season to an end, is associated with the Nativity (El Niño is Spanish for the Christ Child). In some years, the warming is much stronger and longer than usual and prevents the upwelling of cold, nutrient-rich waters which sustain the fish stocks. These above-normal SSTs are part of a major

Figure 4.7. Sea-surface temperature anomalies (°C) during a typical ENSO event obtained by averaging the events between 1950 and 1973. The progression shows (a) March, April and May after the onset of the event; (b) the following August, September, October; (c) the following December, January and February; and (d) the declining phase May, June and July more than a year after the onset. (From Philander, 1983. With permission of Macmillan Magazines Ltd.)

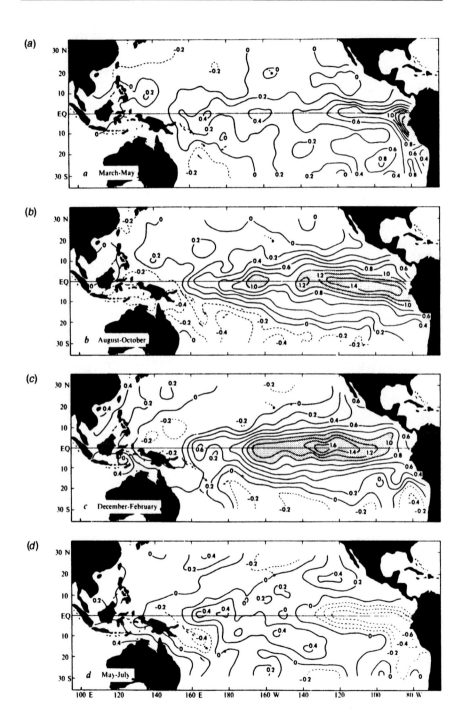

adjustment of temperature patterns all the way across the Pacific (Fig. 4.7). Because it is now recognised that such El Niño events are linked with the behaviour of the Southern Oscillation, the two are now considered together as ENSO events.

The implications of ENSO events for tropical weather patterns are now well established. As the equatorial Pacific warms, the area of heavy rainfall over Indonesia moves eastwards to the central Pacific. At the same time rainfall over Australia declines, and during major ENSO events, such as the one in 1982–83, virtually the whole of the continent is afflicted by severe drought. Over South America the shifts are less pronounced but significant. The area of heaviest rainfall over Amazonia moves to the west of the Andes, bringing torrential falls to the coastal regions of Ecuador and northern Peru. Over much of the Indian subcontinent, ENSO events are associated with a weakening of the monsoon. But it is over Africa that the most complicated developments take place. The area of ascending air over equatorial regions of the continent tends to be replaced by descending motion. This change has been linked with the widespread drought in sub-Saharan Africa in recent decades. The patterns of drought cannot be linked in a simple way to ENSO events and so need to be examined in more detail. What is clear, however, is that it is Africa that stands to be a major beneficiary of greater understanding of how the atmosphere and oceans combine to produce major interannual and interdecadal changes in the climate.

The reason for this is that more than almost any other event in the developing world, drought in Africa has come to represent both the threat of global warming and the pressure of growing populations on limited natural resources. The images from the Sahel in the early 1970s when over 100 000 people died (Fig. 4.8) or the Ethiopian famine in 1984, which had even greater mortality, brought home to many people in the developed world the human consequences of climatic change. At the same time, the recurrent droughts in Africa have provided some of the most formidable evidence of how the climate throughout the tropics is driven by the fluctuations in SSTs and, in particular, of the ENSO.

The drought in the Sahel started in earnest in 1968 (see Fig. 1.3) and reached its first peak in 1972. It abated to a certain extent but returned in the late 1970s and with greater vigour during the 1980s before easing off during the 1990s. The impact of this prolonged example of climatic change has to be viewed against the background of the climatology of the Sahel. This semi-arid region stretches in a narrow band across Africa from

Figure 4.8. An example of the tragic consequences of the drought of the early 1970s in the Sahel. (Reproduced by permission of Hulton Getty.)

Senegal and Mauritania to the Red Sea. Rainfall amounts vary from 200 mm to 800 mm a year from the north to south of the band, and are restricted to the rainy season from June to September. These amounts vary appreciably from year to year and are linked with the precise position of the northward movement of the fringe of the region of strong convective activity known as the intertropical convergence zone (ITCZ). The movement of this zone, which girdles the globe, is controlled principally by the annual tracking of the Sun north and south of the equator, but its strength and precise position is influenced by tropical SSTs. Thus Sahel rainfall is by nature erratic and subject to remote forces. Indeed, historical studies indicate that severe droughts occurred in the 1680s, 1740s and 1750s, and 1820s and 1830s.[20]

The sustained nature of the drought in the Sahel means that it cannot be attributed solely to links with the ENSO, which varies too much on a timescale of a few years. Initial speculation in the 1970s about a permanent shift in the ITCZ as a product of desertification, in part resulting from overgrazing by herds belonging to nomads in the region, has not been confirmed. Moreover, an argument gained wide currency that the change in the amount of sunlight reflected back to space (the albedo) would produce a positive feedback which would reinforce the process of desertification. Because desert sand reflects much more sunlight than vegetation the amount of heat absorbed by a desert would be correspondingly less; this would reduce convection and enfeeble the rainy season – so once the desert was created it would remain. This has not been substantiated. Indeed the capacity of the vegetation to regenerate in wetter years has been stikingly confirmed by satellite observations.[21] So the explanation will probably be found in wider climatic patterns associated with changes in SSTs.

The wider issue of deserts advancing caught on in the 1970s because of events in the Sahel. Terrifying figures entered environmental mythology. More than 20 million hectares (an area well over half the size of the British Isles or the size of Kansas) of once-productive soil was being reduced to unproductive desert each year. The image of the Sahara marching inexorably southwards at up to 50 km a year galvanised many aid agencies into action. This concern culminated in the UN Conference on Desertification, held in Nairobi in 1977. It launched a plan of action which funded projects amounting to some $6 billion over the subsequent 15 years to prevent desertification. During the same period increasing doubts have arisen as to whether the whole concept of desertification was misconceived and what was really needed were better measures of the changes that were actually occurring. In particular, there was no adequate distinction between degradation due to human activities (e.g. overgrazing by pastoralists' herds, collection of firewood, and inappropriate farming) and the effects of drought.[22]

Data collected in recent years suggests that the dominant role of climatic change had been underestimated. Many of the observed shifts in the desert were in fact largely due to annual fluctuations in rainfall.[21] Furthermore, together with the satellite observations of the re-establishment of vegetation in wetter years, more detailed studies at local level showed an astonishing capacity for what looks like complete desert to spring to life when heavy rain falls and long-dormant seeds germinate. This contradicts the implicit assumption of many environmentalists in the 1970s that not only were the changes the result of human activities, but also that they were irreversible. This latter point is central to much of the debate of climatic change. There is a risk of underestimating the capacity of many forms of life to adapt to sudden and massive changes in the climate. We should not forget that they evolved through periods of far greater climatic variability (see Section 5.1) than have been experienced in recorded history. This means that they contain genetic defences which enable some of their species to survive through a wide variety of extremes.

Elsewhere in Africa droughts have been more erratic and have not shown the sustained behaviour of the Sahel events. The links with the ENSO are, however, more definite. Warm El Niño events are linked with below average rainfall in southern Africa. In 1991–92 the start of the prolonged ENSO event was associated with the worst drought in southern Africa this century, affecting nearly 100 million people. Work between the Lamont Doherty Observatory and the Southern Africa Development Community (SADC) Food Security Technical and Administrative Unit in Harare, Zimbabwe has shown how important these links are.[22] The correlation coefficient[23] between the SST in the eastern tropical Pacific and both annual rainfall variation and maize yields in Zimbabwe are extraordinarily high. The figure for the period 1970–93 for the SST-rainfall link is +0.64, while the SST-maize yield is even higher at +0.78. The difference between the rainfall and yield figures have only a one-in-five probability of being the product of chance, which suggests that the yields amplify rainfall fluctuations in the drier parts of the country.

Predicting the temperature in the tropical Pacific is one of the most successful areas of seasonal forecasting (see Section 6.1) and provides a powerful indication of the potential benefits of these predictions. Deceived by good rains in October and November 1991, the governments of the SADC were not prepared for the severe drought that followed. Because grain stocks were already low, some 11.6 million tons of drought-related commodities had to be imported within a 13-month period. Although widespread famine was averted by this action, it was costly and precarious. Timely forecasts might have averted some of the costs, and the fear of worse consequences.

The effective management of emergency grain supplies is likely to become a much more demanding process in the future. World grain stocks have declined sharply since the late 1980s as demand has risen and production has stagnated. By the spring of 1996 they had sunk to well below 50 days' consumption: the lowest level since 1948, down from over 100 days in the mid-1980s. These changes have been the product of a combination of rising per capita consumption in increasingly affluent countries like China, the reduction of areas planted by major producers (e.g. the EU and the USA), warfare, and population increases. Adverse weather was not a major factor in this decline, and the markets took these developments in their stride until the combination of a cold winter and late spring sowings led to gloomy forecasts of the US wheat crop for 1996. Prices rocketed in late April and briefly were some 50 per cent above 1995 prices, reaching levels not seen since 1973, following the 'Great Grain Robbery' (see Section 4.4). While this upsurge in prices was shortlived, as both the harvest in North America and plantings in the southern hemisphere were above expectations, it demonstrated yet again that it is only when stock levels fall below a critical level that the impact of bad weather really starts to be felt.

4.6 Summary

All these examples provide ample evidence of the disruptive and costly consequences of extreme weather events. The next question is how much of this impact could have been alleviated if there had been better forecasts. But before we consider the potential for modelling the climate and economic systems, we need to put the selection of extreme events discussed so far into the context of the wider knowledge of how the climate has changed and what this means for the future.

4.7 Notes

1 IPCC (1995), p. 334.
2 Landsea *et al.* (1994).
3 Landsea (1993).
4 Nutter (1994).
5 Meteorological Office (1987).
6 McCallum (1990).
7 Hulme & Jones (1991).
8 WMO (1995), p. 78.
9 Ibid., p. 76.
10 Lott (1994).
11 Fink, Ulbrich & Engel (1996).
12 See Chapter 10 by Mitchell & Ericksen in Mintzer (1992).
13 Pimentl *et al.* (1995).
14 Kates, Ausubel & Berberian (1985), p. 373.
15 Parker, Legg & Folland (1992).

16 Wigley, Lough & Jones (1984).

17 Kates, Ausubel & Berberain (1985), p. 92.

18 Lamb (1972), pp. 240–50.

19 Philander (1983).

20 See Chapter 9 by S. E. Nicolson in Wigley, Ingram and Farmer (1981).

21 Tucker, Dregne & Newcomb (1991).

22 Thomas & Middleton (1994).

23 Cane, Eshel & Buckland (1994).

24 This is a mathematical estimate of the linear association of two variables. The closer it is to +1 the closer the association.

5

How much do we know about climatic change?

'It's a poor sort of memory that only works
backwards,' the Queen remarked.
Through the Looking-Glass, Chapter 5

In describing the most significant examples of the economic impact of climatic change, a certain amount of information has been given about the overall nature of climatic change during the last millennium. By concentrating on the most dramatic events there is, however, a risk of presenting a partial picture of what happened. So it is important to have a balanced view of the current state of knowledge about past changes. Only then will we be able not only to put the ups and downs of the past into context but also start to form a view as to whether this knowledge can be applied in preparing for the future. In particular, without a better understanding of the extent of past natural climatic change it is not realistic to plan on the basis that current changes are the consequence of human activities. But, before considering the evidence of climatic change in recent centuries, we need to think the unthinkable.

5.1 Chaos round the corner?

The quest for a better understanding of the economic impact of climatic change on our current world has concentrated on relatively recent events. The far greater changes that have occurred over geologic timescales seem far too remote to bother about here. The forecast that, in the absence of human activities, the Earth will slip back into the next ice age in about 23 000 years seems wholly irrelevant.[1] Recent research has, however, cast particular doubt on the cosy notion that vast changes in the climate on the scale of the ice ages occur with glacial gradualness.[2]

The essence of the new studies is that the climate has been extraordinarily stable for the last 10 000 years. Prior to this, as the Earth emerged from the last ice age and throughout the preceding glacial epoch the climate was much more erratic. Even more dramatic is some evidence that during the previous interglacial (the Eemian, 115 000 to 135 000 years ago, when the global climate was considerably warmer than now) there were sudden and substantial shifts in the climate. Whereas the shifts we have been considering so far are of the order of, at most, one degree Celsius over a few decades or possibly centuries, the earlier changes were five to ten times as great and occurred over a few years.

The principal source of evidence of the more erratic climate before 10 000 years ago are recent ice cores taken from the Greenland ice sheet.[3] Working at an altitude of 3200 metres in the icy wilderness on the top of the ice cap (Fig. 5.1), scientists have extracted two three-kilometre long vertical cores from the ice which provide a picture of climatic fluctuations over the last 150 000 years. Measurement of various properties of the ice tells us a great deal about the climate of the past. Shifts in the ratios of the stable isotopes of hydrogen and oxygen provide information about changes in temperature from year to year. The amount of dust is an indication of wind strengths, while carbon dioxide and methane concentrations trapped in air bubbles give clues as to the causes of changing climate, and sudden jumps in acidity record the timing and size of major volcanoes. What these measurements show is that the first 1500 metres of the cores, containing snow that fell over the past 10 000 years, shows a relatively stable climate (Fig. 5.2). While there are significant shifts, including the cold period that wiped out the Norse settlements in Greenland (see Section 2.1), by comparison with what preceded it, the climate has been rock steady.

Deeper down, a more dramatic story emerges. By counting individual layers of snow back to 15 000 years ago, climatologists have been able to observe the erratic climb out of the last ice age. While the broad pattern of warming and the sharp reversal between 13 000 and 11 600 years ago confirmed earlier work, the existence of enormous fluctuations in temperature and volume of snowfall within a few years came as a surprise. Prior to this time, it is not possible to measure individual layers, but the picture

Figure 5.1. A rig drilling an ice core at Vostok on the Antarctic ice sheet. (Reproduced by permission of L. Augustin, CNRS.)

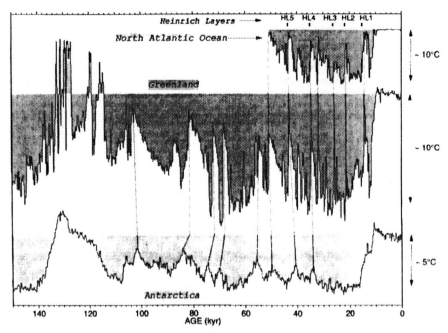

Figure 5.2. Reconstructed climate records obtained from ocean sediment faunal records in the North Atlantic and analysis of isotope levels ice core records from Greenland and Antarctica for last 150 000 years, showing how major changes in sediment-formation (Heinrich layers) are linked to the sudden rises in the temperature inferred from the ice cores. (From IPCC, 1995, Figure 3.22.)

of an erratic climate prevails throughout the whole of the last ice age back to some 100 000 years ago.

A far more startling discovery, in one ice core,[4] was the evidence of an erratic climate during the Eemian. Before these observations it had been assumed that this interglacial had a climate similar to that of today, and that relative stability was the norm for interglacial periods. Instead, evidence emerged of three markedly different climatic regimes: one was much colder than the present, one was similar to today, and one markedly warmer. Most striking was the scale of the changes between these regimes. On some occasions, within a decade or so, the average temperature shifted by 10 °C or more and then remained constant for anything from 70 to 5000 years. Tantalisingly, these observations were not confirmed by the second ice core, drilled 30 kilometres away.[5] There is now a debate within the climatological community as to which set of observations is closer to reality. The matter will not be resolved until further cores are drilled and the evidence is found to stack up one way or the other. In the meantime other more readily available sources of climatic information for the Eemian are being explored or re-examined to see whether they support the stable or erratic interglacial model. But it is the nature of this type of debate that not only are data from elsewhere in the northern hemisphere equivocal but also there is the complicated issue of whether changes observed over Greenland are representative of the rest of the world.

Whatever the outcome of future research, the Greenland ice cores have had a major impact on climatic thinking. Whether or not the Eemian turns out to have been stable or erratic, the existence of sudden large shifts in temperature and precipitation have altered perceptions on the pace at which the climate changes. Termed a 'flickering switch' by researchers, these ups and downs were much larger and more rapid than anything in recent experience.[6] It can be argued that during both the last ice age and the warming that took place around 12 000 to 14 000 years ago the conditions may have been inherently less stable owing, in particular, to sudden collapses of the North American ice sheet flooding the Atlantic with icebergs – known as Heinrich events[7] (see Fig. 5.2). Nevertheless, the capacity of the climate to shift so suddenly has led to a whole new set of models of how the oceans might be capable of switching between different circulation patterns. So, whereas earlier thinking had assumed that the thermal inertia of the oceans meant that big changes in the climate took a long time, now it is possible to conceive of much more rapid and chaotic shifts.

The possibility of the oceans participating in sudden and huge shifts in the climate is a spectre that lurks behind all efforts to predict the future of global warming (see Section 5.8). It takes us into the world of Chaos Theory[8] and the inherent unpredictability of non-linear systems. While it is possible to speculate on possible shifts in the circulation of the oceans and what this might do to the climate, we have no basis for asserting that any particular scheme is likely to occur. This means that in examining the more modest aspects of recent climatic change and how this should influence our thinking about predictions of global warming and its economic impact, we may discover that like the drunk under the lamplight we have been searching where the light is best and not necessarily where the lost key is to be found.

The existence of such rapid changes also touches on another important issue. This is the capacity of flora and fauna to adapt to rapid changes in the climate. Whereas sudden large shifts in the climate would inevitably pose immense challenges for human activities, it has also been assumed that the ecological impact would be equally devastating. Much has been made of the fact that predicted temperature rises in the coming decades will be far more rapid than anything seen in the past, notably during the emergence from the last ice age. But if much more rapid changes in the climate were the norm prior to 10 000 years ago (at least during the last million years or so which have featured periodic ice ages), it follows that flora and fauna have evolved to adapt to such an erratic climate.

5.2 The last 10 000 years

Against the uncertain background of the chaotic climate associated with the last ice age, it is wise to put the subsequent stability in context. As Fig. 5.2 shows, the climate of the last 10 000 years has been relatively stable, but hidden within this orderly picture are the significant developments which underlay the events described in Chapters 2 to 4. Initially the relative stability observed in Greenland would not have been reflected in other parts of the world. As the world adjusted to the collapse of the massive ice sheets over North America and Scandinavia, the influxes of fresh water into the North Atlantic and the rise of sea level would have had major impact on regional climates. But by around 7000 years ago the climate reached an optimum. By this time the temperature in mid-latitudes of the northern hemisphere were around 2 °C warmer in summer than conditions during the twentieth century (Fig. 5.3). The strength of the summer monsoon was also greater, bringing more abundant rainfall not only to the Indian subcontinent but also to the Sahara.

Around 5500 years ago a gradual cooling trend appears to have set in. At the same time the monsoon circulation in the subtropics started to weaken and increasing desiccation began to affect the Sahara in particular. But it was not until around 4000 years ago that a more marked shift occurred. In the Middle East and North Africa there is considerable evidence of a decline in rainfall, with Egypt experiencing a 'dark age' between the fall of the Old Kingdom around 2200 BC and the emergence of the Middle Kingdom in about 2000 BC. Across the Canadian Arctic and Siberia cooling led to the northern extent of the tree line receding some 200 to 300 kilometres.

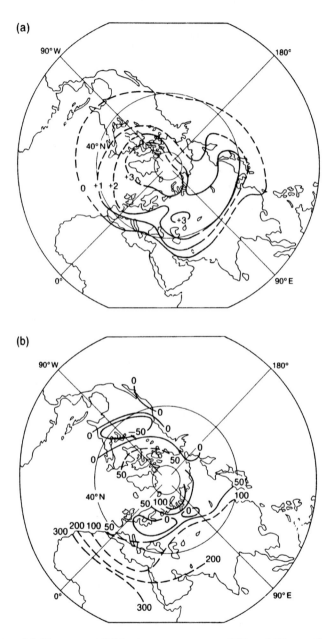

Figure 5.3. Departures of: (a) summer temperature (°C), and (b) annual precipitation (mm), from modern values for the Holocene climatic optimum around 6000 years ago. (From Borzenkova & Zubakov, 1984; Budyko & Izrael, 1987.)

Thereafter the gradual global cooling and desiccation in the subtropics continued until around 1000 AD. Superimposed on this trend was a series of wiggles that brought warmer and wetter periods interspersed with cooler drier episodes. What is not clear is how rapid and intense these fluctuations were. So, as noted in Chapter 2, this uncertainty leaves plenty of room for argument about the scale and consequences of climatic change in recorded history. But, as a general observation, there is widespread evidence of glaciers expanding in mountains at high latitudes around 3000 to 2500 years ago and again between around 500 and 800 AD. Depending on how these changes are translated in global atmospheric circulation patterns, they can be used to support the various theses about changes in rainfall regimes at lower latitudes may lurk behind the unexplained decline in ancient civilisations discussed in Chapter 2.

5.3 Medieval climatic optimum

Moving towards the present, we can draw on many more records, and so the evidence of climatic change increases. Although these records give only a partial picture of global fluctuations, they provide a clear picture of a marked warming in Europe and around the North Atlantic during the ninth and tenth centuries. This period coincides with the Norse colonisation of first Iceland and then Greenland. The records of agricultural activity and fishing suggest that by around 1000 AD the sea surface temperatures in the north-west Atlantic were comparable to the warmest values recorded during the twentieth century.[9]

In Europe a similar picture emerges. The advancing of agriculture to higher altitudes in places such as Norway and Scotland suggest warmer summer conditions. Also the success of wine production in England is widely seen as proof of a milder climate. More generally the expansion of agriculture, commerce, cultural activity and population all point to the same conclusion. So, although the reports of agriculture are still punctuated with poor harvests and occasional sharp price rises, the eleventh and twelfth centuries present a general picture of expanding economic activity consistent with an overall improvement in the climate. Broadly speaking it is estimated that the summer temperature in north-western Europe was comparable to or a little warmer than figures for the twentieth century (Fig. 5.4).

Tree-ring data from northern Fennoscandia[10] and the Urals,[11] which

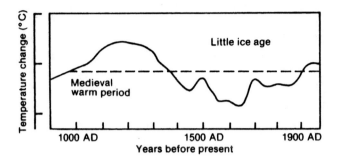

Figure 5.4. A schematic diagram of global temperature variations over the last 1000 years (solid line) as compared with conditions at the beginning of the twentieth century (dashed line). (From IPCC, 1990, Figure 7.1(c).)

have been analysed to highlight longer term fluctuations in summer temperatures, provide an independent check on the inferences drawn from historic records. In northern Fennoscandia there is evidence of a warm period around 870 to 1110 which coincided with the European medieval climatic optimum. These data also show a warm period around 1360 to 1570. The records for the northern Urals show a somewhat different picture. There is no evidence of marked warmth in the tenth and eleventh centuries. Instead it is the thirteenth and fourteenth centuries together with the late fifteenth century that show the most warming, suggesting shifts in circulation patterns rather than changes in hemispheric temperature levels.

These regional variations also emerge from other historical records. As noted in Chapter 2 the cooling in Greenland set in considerably earlier than in Europe. So we are confronted with fragments of the records which suggest that synchronous and substantial shifts in global climate cannot be used to explain much of the longer term variability identified in available records. The changes may be part of altered patterns of global circulation which brought sustained but compensating periods of abnormal weather to different parts of the northern hemisphere. But, given the wide variety of different patterns that could have been established, we cannot make assumptions about what conditions prevailed in regions where we have no reliable data.

As explained in Chapter 2 the medieval climatic optimum in Europe appears to have gone into decline during the thirteenth century. Thereafter the climate appears to have been less clement throughout Europe.

But there is relatively little clear-cut evidence of appreciable further decline during the fourteenth and fifteenth centuries. Rather, the picture is of greater variability on the interannual and interdecadal scale.[12] Also, if anything the climate warmed during the fifteenth century, and this warmth extended up to around 1550. The tree ring records from northern Fennoscandia provide a similar chronology. The Chinese records for the period suggest a cooling in the eastern part of the country during the fifteenth century, but if anything a warming in the north.[13] Again we see a patchy combination which provides little indication as to whether any pronounced global changes were in train.

5.4 The Little Ice Age

By comparison with the uncertainties of preceding centuries, the evidence of the cooler period between the mid sixteenth and mid nineteenth centuries appears to be built on firmer foundations. It is the best known example of climatic variability in recorded history. The popular image is of frequent cold winters with the Thames in London being frozen so that Frost Fairs could be held on the ice. Elsewhere in Europe the same image of bitter winters prevails, together with periods when cold wet summers destroyed harvests as described in Chapter 2. Widely known as the 'Little Ice Age', the period has been closely studied by climatologists for many years. This growing body of work shows that, as with all aspects of climatic change, the real situation is not quite as stark as the simple stereotype suggests.

There is no doubt that the climate in Europe deteriorated in the second half of the sixteenth century. The glaciers in the Alps expanded dramatically and reached advanced stages at the end of the 1590s. The work of Christian Pfister clearly shows that in Switzerland the period 1570 to 1600 featured an exceptional number of cool wet summers (see Fig. 2.9).[14] This work also confirms that the winters of the 1590s were particularly severe. The Ladurie wine harvest dates provide confirmation of the poor summers of the late sixteenth century (see Fig. 2.6).[15]

Where the interpretation of the Little Ice Age as a period of sustained cold breaks down is in the subsequent decades. Both Pfister and Ladurie's work shows considerable fluctuations from year to year. But, for the growing season, the cold of the 1590s is not maintained. There is, however, evidence of more frequent cold winters in the commercial records of

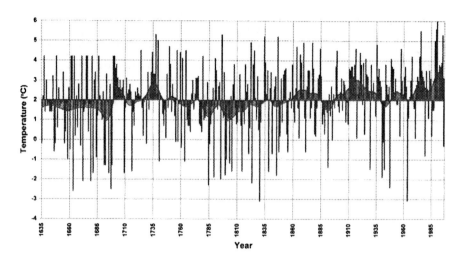

Figure 5.5. The winter temperature record for De Bilt in the Netherlands since 1634, together with smoothed data showing longer term fluctuations. (Data taken from Van den Dool *et al.*, 1978, and Engelen & Nellestijn, 1995.)

Dutch merchants which noted when the canals were frozen and trade interrupted. These show that from 1634 until the end of the seventeenth century the winters were roughly 0.5 °C colder than in subsequent centuries.[16] Most striking is the particularly cold winters of the 1690s (Fig. 5.5).

The cold wet nature of the 1690s can be found in nearly all the available records. Only in the wine harvest records and the Swiss summer records does it not show up as the outstanding feature of the European climate in the last 500 years. This suggests that a cool dry summer did less harm to the grapes than to other crops. Otherwise, the 1690s stand out in Swiss seasonal data, with the figures for spring (see Fig. 2.8) standing out most dramatically, showing that the Alps suffered particularly from cold snowy conditions which retarded the growing season substantially.

By now we have the first of the instrumental records (see Section 2.4) with the Central England Temperature (CET) series starting in 1659[17] and the record for De Bilt in the Netherlands in 1705 (this record[18] is combined with the canal freezing data to provide a winter series for the station since 1634 in Fig. 5.5). The annual figures for the CET series confirms the exceptionally low temperatures of the 1690s (Fig. 5.6) and, in particular, the cold late springs of this decade. The first striking feature of these records is the sudden warming from the 1690s to the 1730s. In less than 40 years the conditions went from the depths of the Little Ice

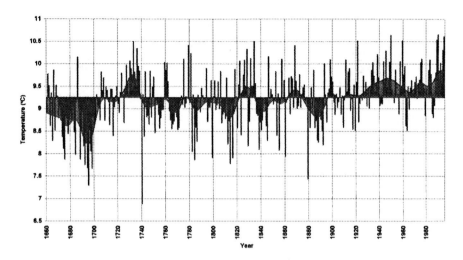

Figure 5.6. The Central England Temperature record showing the variation
in annual temperatures between 1660 and 1995, together with smoothed data
showing longer term fluctuations. (Data from Manley, 1974, and Parker *et
al.*, 1992.)

Age to something comparable to the warmest decades of the twentieth
century. This balmy period came to a sudden halt with the extreme cold
of 1740 and a return to colder conditions, especially in the winter half of
the year. Thereafter the next 150 years or so do not show a pronounced
trend. It is a feature of not only the CET and De Bilt series but also the
various other series that have been found for other European cities from
the mid eighteenth century onwards. Although the winters begin to
become noticeably warmer from around 1850, the annual figures did not
show any appreciable rise until well into the 20th century. Furthermore
the temperatures for the growing season (April to September) remained
virtually steady until the present day. The emergence from the Little Ice
Age, such as it was, in Europe was concentrated in the winter half of the
year.

   What is more striking in the temperature series, and for that matter the
proxy records, is the interdecadal variability. The cold wet summers of
the 1810s (see Section 2.4) are incontrovertible, as are the hot ones of the
late 1770s and early 1780s, and those around 1800 are equally prominent.
The same interdecadal variability shows up in more recent data. The
1880s and 1890s were marked by more frequent cold winters, while the
1880s and 1910s had more than their fair share of cool wet summers.

A similar variable story emerges from other parts of the world. In eastern Asia the seventeenth century was the coldest period, with another set of cold decades around 1800. But the rest of the nineteenth century does not show the frequent periods with low temperatures seen in Europe.[19] Conversely, the North American records show that the coldest conditions were in the nineteenth century. Tree-ring data suggest that the seventeenth century was also cold in northern regions, but in the western United States this period appears to have been warmer than the twentieth century. Limited records for the southern hemisphere indicate that the most marked cool episodes occurred earlier, principally in the sixteenth and seventeenth centuries.

In summing up the evidence of the Little Ice Age, Bradley & Jones[20] reach a number of conclusions. Most significant is that the last 500 years have not experienced a monotonously cold period: certain intervals have been colder than others. Only a few short cool episodes appear to have been synchronous on the hemisphere and global scale. These are the decades 1590s to 1610s, the 1690s to 1710s, and 1800s and 1810s and the 1880s to 1900s. Synchronous warm periods are less evident, with the 1650s, 1730s, 1820s, 1930s and 1940s being the most striking. The lack of synchronicity means there is geographical variability in climatic anomalies, the coldest episodes in one region often not being coincident with those in other regions. Finally, as noted in the case of Europe, different seasons often show different anomaly patterns over time. Overall, this spotty picture led Bradley & Jones to conclude that the term 'Little Ice Age' should be used with caution.[20]

This guarded conclusion to what is widely seen as the 'most relevant examples of past climatic change in recorded history', provides further confirmation of the need to concentrate on regional patterns of change (see Section 1.5). Sudden and substantial changes in any country will have immediate political consequences irrespective of what is happening globally. Conversely, the absence of any appreciable trend in local conditions will emasculate any resolve to make sacrifices to address global problems. So it is essential that we understand the regional structure of past climatic change in assessing the threat of future developments.

5.5 The twentieth century warming

Central to all discussions on the impact of human activities on the climate is the warming during the twentieth century. Not only is it seen by many

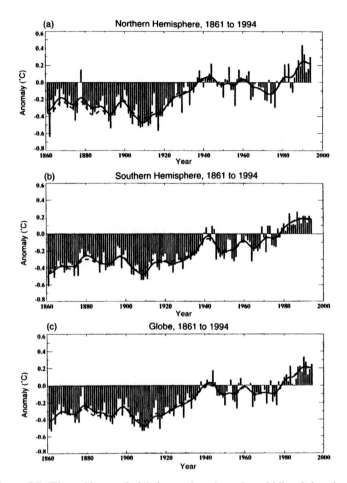

Figure 5.7. The evidence of global warming since the middle of the nineteenth century can be seen in the combined annual surface air and sea surface
temperature (°C) anomalies relative to the 1961–90 average (bars and solid
smoothed curves) for (a) the northern hemisphere, (b) the southern hemisphere and (c) the entire globe. The dashed curves are earlier estimates published in IPCC (1992). (From IPCC, 1995, Figure 3.3.)

as evidence of anthropogenic effects but also it is the yardstick against
that to test the models which seek to quantify the impact of these effects.
The changes for the northern and southern hemisphere that have occurred
since the late nineteenth century are shown in Fig. 5.7. The important
features are that there has been a general warming of between 0.3 and

0.6 °C, and that most of this warming has been concentrated in the period 1920 to 1940 and since the mid-1970s.

Within this broad upward trend there are a number of features which need to be addressed here. There is the fundamental question of whether the changes are an accurate measure of what the global climate is doing or are nothing more than artefacts of changes in measurement techniques, in movement of observation sites between urban and rural areas, and in the geographical coverage around the world (the northern land masses have been well covered, while there are very few data for the southern oceans). Needless to say, these issues have been explored in great detail to cover such questions as the standardisation of the measurement techniques, the correction needed for the effects of urbanisation, and how the water was collected for observations of sea surface temperatures (an essential element in getting adequate global coverage) – for example, whether canvas or wooden buckets were used for the early measurements.[21] The figures shown in Fig. 5.7 reflect the results of this intensive scrutiny of the data, and the Intergovernmental Panel on Climatic Change (IPCC) reports discuss the outcome of the analysis. While the conclusion is that by far the greatest part of the observed temperature rise is real, there are still some interesting questions to be answered about the best correction for the warming due to urbanisation. There is also some uncertainty about the exact levels of the temperature in the earliest part of the record and whether they have an impact on how this much-researched series dovetails into the analysis of earlier changes.

One obvious feature in these temperature trends, which must be explained by climate modellers, is the slight cooling between 1940 and 1970 in the northern hemisphere. If the warming trend since the late nineteenth century is due to the build up in greenhouse gases resulting from human activities, then some additional agency needs to be invoked to explain this setback and the subsequent sharp rise in temperature. The next level of sophistication for the modellers is to simulate the regional features of the recent warming trend. Analysis of the warming in the 1980s shows that there are distinct variations on a regional scale in the annual figures and in each season. Most notable is pronounced warming over much of the northern continents, a marked cooling in the north-west Atlantic and a less pronounced cooling across the central northern Pacific. These changes show up most dramatically in the winter – December to February (Fig. 5.8). The reasons for this pattern are still the subject of scientific investigation, and, as such, constitute an obvious challenge for

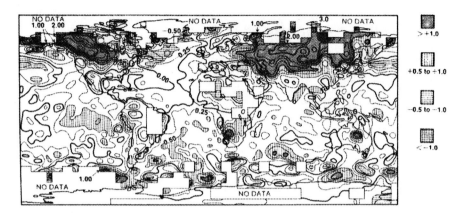

Figure 5.8. Changes in winter (December to February) temperature during the 1980s, compared with the average figures for 1951–80 showing that the most marked warming occurred at high latitudes, but that there were areas of cooling notably in the vicinity of southern Greenland. (From IPCC, 1992, Figure C.5(b).)

modellers in showing that the current warming is linked to human activities. This pattern is suggestive of what is known as the North Atlantic Oscillation, which will be discussed in terms of understanding the natural variability of the climate (see Section 5.7). Clearly, it is essential to establish how much of these recent changes can only be explained in terms of human activities and to what extent they are the product of natural feed-

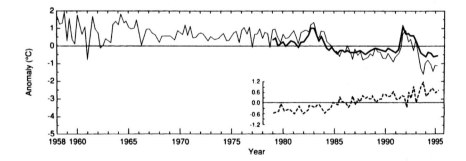

Figure 5.9. The cooling of the stratosphere between altitudes of 16 km and 21 km observed by radiosondes (thin line) and satellites (thick line), together with the difference between the two sets of observations (dashed line). (From IPCC, 1995, Figure 3.7.)

Figure 5.10. Large volcanic eruptions can have a significant impact on the climate. This is the eruption of Mount St. Helens in 1980. (Reproduced by permission of US Geological Survey, Cascades Volcano Observatory.)

back mechanisms in the climate. Given how, in the past, there appear to have been appreciable regional variations in climatic change, this is a fundamental issue.

The same challenge relates to how the models handle changes in the stratosphere. Regular measurements have been sufficiently widespread since the early 1960s to show that there has been a marked cooling of the stratosphere during the last 30 years (Fig. 5.9). Although this is a relatively short record, the observed trend is consistent with the model predictions but the scale of the cooling is on the high side. This difference may reflect, however, problems with the measurements rather than with the models. Changes in the design of balloon-borne instrument packages (radiosondes) have, over the years, reduced the chance of temperature readings being affected by sunlight. So part of the cooling may simply be a product of improved equipment design.

Another complication in the stratosphere is the impact of major volcanoes (Figs. 5.10 and 5.11). The injection of large quantities of dust and sulphuric acid aerosols into the upper atmosphere leads to significant absorption of sunlight in the stratosphere. Warming at these levels follows

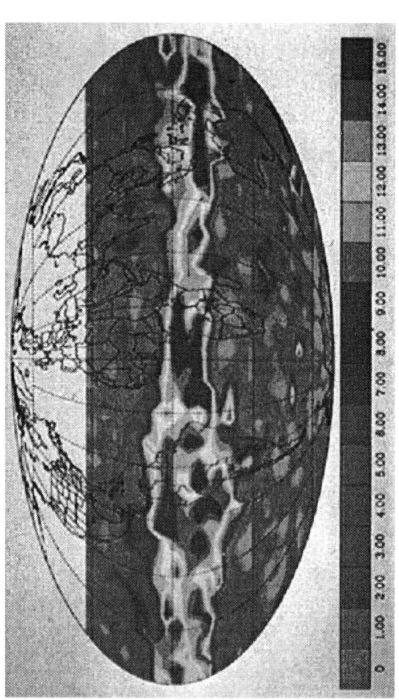

Figure 5.11. Satellite measurements of the spread of the dust cloud from the eruption of Pinatubo in June 1991, showing that within five weeks the dust girdled the globe. (With permission of NOAA.)

and a corresponding but smaller cooling occurs lower down. Three major eruptions (Agung in 1963, El Chichón in 1982, and Pinatubo in 1991) can clearly be seen in the record and further complicates the interpretation of the cooling trend over the last 30 years or so. But, as we will see later, the computer models handle the impact of volcanoes rather well (see Section 6.2).

The other measurement issue is an ongoing debate about the different results obtained from the surface observations and data collected by orbiting weather satellites. Because satellites can provide genuinely global coverage of the atmosphere they are, in principal, the ideal platform for monitoring global temperatures. The technique they can use relies on the fact that the amount of sunlight absorbed in the Earth's atmosphere and at its surface, and the amount of heat radiated back into space, are in balance. Because the amount of heat radiated depends on the temperature of the atmosphere and the surface beneath, it is possible to make accurate observations of the Earth's temperature by measuring the outgoing for any part of the globe. To do this, weather satellites use radiometers to scan the Earth's surface and detect the tiny amounts of energy radiated at different infrared and microwave wavelengths.

Using infrared radiometers it is possible to make independent observations at four levels in the atmosphere below an altitude of ten kilometres. Microwave radiometers are able to resolve only two levels – effectively the troposphere, centred on an altitude of 3 km (10 000 ft), and the lower stratosphere – but have the huge advantage they can see through clouds. Because clouds are efficient emitters of infrared radiation they play havoc with temperature measurements by infrared radiometers. Microwave radiometers have only been flown in weather satellites since 1979. Using measurements made by these instruments, Roy Spencer, of the NASA Marshall Space Flight Center, Alabama, and John Christy of the University of Alabama, have produced a 16-year series of changes in the temperature of the lower atmosphere.[22]

The intriguing feature of this work is that, unlike the ground-based observations, the satellite record shows a much smaller warming trend since 1979. This discrepancy has led to much head-scratching. In principle, the combination of the sensitivity of the satellite radiometers and the truly global coverage should produce more accurate measurements. Furthermore where the measurements from space can be compared with extensive ground-based readings, as over the USA the agreement between the observed fluctuations in temperature is exceedingly good. The cause of the discrepancy is the subject of a continuing debate. James Hansen and colleagues at the Goddard Institute of Space Studies, New York, have concluded[23] that the difference in surface and satellite temperature trends (at least 0.12 to 0.15 °C over 15 years) is almost certainly significant, but can be explained in terms of a combination of factors. These relate to the fact that surface trends may be slightly different from the whole of the troposphere, and there may be some small distortions due to variations in the microwave emissivity of the Earth's surface and to water droplets and ice particles in clouds. Possibly more important is the proposal that the decline in stratospheric ozone will lead to cooling of the upper troposphere which can explain nearly half the discrepancy. Examination of the latitudinal variations of the differences shows, however, there is close agreement in surface and satellite observations in mid-latitudes of the northern hemisphere but a discrepancy of over 0.1 °C per decade in the tropics. This suggests that there is an, as yet, unidentified process at work in the lower levels of the tropical atmosphere which must be pinned down, if we are to remove doubts about current temperature trends.

5.6 Greater extremes?

Because so much of the economic impact of the weather is associated with extreme events, it is essential to have some measure of how the incidence of such events has changed in the past and may change in the future. But, as should become clear from the many examples cited in this book, damaging extremes come in many forms, so producing a handy index of trends in damaging weather is easier said than done. Nevertheless, Tom Karl and colleagues at NOAA (the National Climatic Data Center), Ashville, North Carolina, have come up with measures of climatic change for the USA which seek to quantify trends in extreme weather.[24] What is more the indices they have produced may give some idea of how global warming might alter the incidence of these extremes in the future.

The method used is to measure what proportion of the total area of the conterminous USA experienced extremes in either the top or bottom 10 percent of a variety of climatic indicators in any given year. In the case of the US Greenhouse Climate Response Index (GCRI) the indicators were selected on the basis of the concerns of the public and policy-makers and covered increases in temperature, cold season (October to April) precipitation, severe summertime (May to September) drought, proportion of total precipitation derived from extreme one-day events, and decreases in day-to-day temperature variations. This index shows marked peaks in the 1930s and 1950s associated with the widespread droughts in these decades (see Section 2.6) and the marked shift in global circulation since the mid-1970s.

These changes are consistent with the predictions of the effect of global warming. Tom Karl estimates that the probability of this coincidence having occurred by chance is only about one in twenty. The strength of this link does, however, depend to a great extent that global warming will produce a rise in both temperature and precipitation, and especially the proportion falling in extreme (more than 2 inches or 50.5 mm) one-day events. So the changes could be the result of natural variability and tell us little about the sensitivity of the climate system to human activities. But this analysis is a step in the right direction. It suggests that more work should be done in respect of extreme events such as heatwaves, cold waves, unseasonal frosts, high winds, tornadoes and hurricanes.

More generally the IPCC conclusion is that there is no consistent trend in interannual temperature variability in recent decades,[25] and no consistent pattern for rainfall variability. The same story emerges in respect of intense rainfall and extratropical cyclones. Tropical cyclones appear to have declined in the North Atlantic but the observations are not sufficiently reliable to draw any conclusions. Overall, there is no clear evidence that extreme weather events, or climate variability, has increased, in a global sense, during the twentieth century.

5.7 Statistical observations

The problems of analysing the occurrence of extreme weather is a subset of the general problem of not falling into statistical traps when testing the significance of changes in the climate. So, in discussing the nature of the variability of our climate, it helps to know about what the statistics tell us about fluctuations in the weather. The first thing to say is that on every timescale the weather is varying, whether it be from day to day, year to year or century to century. In mid-latitudes, these fluctuations are closely tied to the behaviour of the prevailing westerly winds and the eddies (depressions) in them which continually transport energy from equatorial to polar regions. Variations in the strength and precise course of these westerly winds govern many of the longer term weather patterns in extratropical regions. The course of the winds has the effect that when they get stuck in a meandering pattern which produces, say, a very cold winter over the USA and northern Europe, intermediate regions will have compensating abnormally mild conditions. So on a hemispheric scale the temperature may be close to the climatic normal. On a slightly smaller scale, such patterns can often result in the western USA experiencing one extreme while the other side of the continent is having just the opposite. But, as we shall see shortly, the broad nature of the circulation does influence hemispheric figures in that strong westerly circulation is associated with above-normal temperatures in the winter half of the year, while meandering patterns produce average or below-normal values. In the tropics, the weather tends to be less variable on a day-to-day basis, but the seasonal patterns are closely linked with the behaviour of the tropical oceans.

In mid-latitudes the principal features of the shorter term fluctuations can be summarised in broad rules. First, variability from month to month is much greater than from year to year, and in turn the fluctuations on longer timescales of decades and centuries are correspondingly smaller in amplitude. In the northern hemisphere the fluctuations are usually bigger in the winter half of the year than in the summer. Typically, the difference between, say, the coldest and warmest January in a century-long record will be of the order 10 °C to 15 °C, and between extreme winters over the last few hundred years it is some 7 °C to 10 °C. By way of comparison, most historic rates of long-term change, on both a regional and hemispheric scale are less than 0.1 °C per decade and predictions of future global warming involve trends of around 0.2 °C per decade. So whether due to natural causes or human activities, they are tiny compared with the ups and downs that occur between, say, successive winters in any given part of the world. But, in apparent contradiction to what has just been said, the fluctuations on longer timescales are greater than would be expected if they were purely random. This property, which is often termed 'red noise', reflects the fact that the climate effectively has a 'memory' which is linked with the inertial effects of the slowly varying components of the global climate, notably in the oceans (see Section 5.8).

The second feature is that on all timescales there are what look like regular fluctuations. But critical statistical analysis of them to discover the existence of cycles has proved a largely fruitless activity. Apart from a clear quasi-biennial oscillation (QBO) in the direction and strength of the winds in equatorial stratosphere with a period of about 27 months, and a more shadowy but ubiquitous 20-year cycle which appears in many weather records, there is very little to show for all the effort to find cycles.[26] What is evident, however, is the existence of many quasi-periodic features which are linked to fluctuations in the more slowly varying components of the climate. Associated with the property of red noise in climatic variability, these persistent, but not wholly reliable, features provide important insights into how the global climate functions. They also may hold the key to producing improved forecasts of seasonal weather (see Section 6.3).

A third and related feature is the statistical importance of the timing and frequency of extreme seasons in creating quasi-periodic fluctuations in climatic series. Behind this feature lies the seminal issue of what establishes the persistent weather regimes that are the essential ingredient of these extremes. As noted earlier, the character of the northern hemisphere winter is dominated by the strength and pattern of westerly winds. When this circulation is strong and persistent, much of North American and western Eurasia have very mild winters, while Labrador and southern Greenland have cold weather. Although the strength and form of this circulation pattern varies considerably, in most winters it is its zonal nature which establishes mild conditions over much of the northern continents. Occasionally, however, the norm is interrupted by persistent high-pressure features which produce a meandering pattern. These highs ('blocking anticyclones') form most often in the vicinity of the Greenwich Meridian and in the north-western Pacific. Whereas normally they last from a few days to a couple of weeks, in some winters, like 1947 and 1963, they last many weeks and completely alter the weather. When, as sometimes happens, extremes come at approximately regular intervals (e.g. the cold winters in Europe in 1895, 1917, 1940, 1963, and the group of slightly less cold seasons in 1985, 1986 and 1987) they can give the impression of periodic behaviour.

More important is the changing incidence of these events. As noted earlier, the winter temperatures in Europe have switched between periods when blocking is prevalent to periods when a more pronounced westerly pattern dominates. The early part of the eighteenth century and the first four decades of this century are good examples of when westerlies were more frequent, as has been the case since the late 1980s. Conversely, the frequent cold winters of the late seventeenth, eighteenth and nineteenth centuries are linked to more frequent blocking. The interesting feature of these switches is that they are mirrored by the temperature patterns in Labrador and South Greenland: when it is mild in northern Europe, it is

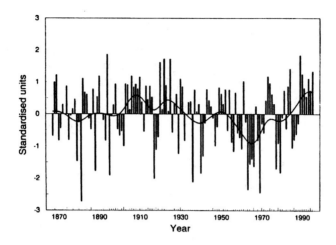

Figure 5.12. The North Atlantic Oscillation as measured by the standardised difference of December to February atmospheric pressure between Ponta Delgada, Azores, and Stykkisholmur Iceland, together with a smoothed curve to show fluctuations longer than around 10 years. (From IPCC, 1995, Figure 3.18.)

unusually cold in these regions. This is not surprising as mild winters in Europe are associated with a persistent deep low in the vicinity of Iceland, which pulls cold arctic air down on its westward flank.

On the other hand, when high pressure develops over Scandinavia and north Europe, mild air is forced up towards Greenland. This switch is known as the North Atlantic Oscillation (NAO), and was first formally described by Sir Gilbert Walker (see Section 7.4).[27] This measure of regional circulation is usually defined in terms of how the pressure difference between Iceland and either Portugal or the Azores varies from the climatic normal each winter. When there is a strong westerly circulation, the pressure difference is large with a deep depression near Iceland for much of the winter and high pressure extending from the Azores to the Mediterranean. A blocked circulation will tend to exhibit the reverse situation with high pressure near Greenland or over Scandinavia and low pressure systems running into the Mediterranean. The fluctuations in the value of this index for each winter since 1867 are shown in Fig. 5.12.

The variations in the NAO are usually discussed in terms of switches over periods of a few years. The longer term shifts in the frequency of cold winters in Europe suggests that this phenomenon may work on the scale of decades and even centuries. What is more, the prevailing wind patterns will tend to establish anomalous SST patterns and of sea ice cover. This in turn could set up ocean circulation conditions which might last for many years following a relatively rapid switch. Although the recorded behaviour of the NAO is nowhere near the scale of past sudden shifts in the region's climate, its behaviour may provide useful insights into both decadal fluctuations in regional climate around the northern hemisphere, and the possibility of more dramatic changes. Furthermore, the sudden switch back to negative values of the NAO in the winter of 1995–96, with a value of −1.25 (cf. Fig. 5.12), gave parts of Scandinavia its coldest November to March period since 1970. The Baltic was frozen solid for three months. Then in December 1996 and January 1997 a similar pattern brought well-below normal temperatures to much of northern Europe: what better reminder could there be for the need to know more about what controls the incidence of circulation regimes?

This brings us back to a fundamental issue. While these observations provide an insight into the nature of the fluctuations in the weather and climatic change, they do not get at the basic question of how the incidence of extreme events will change if the climate warms up. In a nice orderly world, the distribution of a, say, daily temperature would be normal (Gaussian) and then after a period of warming the distribution would shift to a higher mean value while exhibiting the same spread (Fig. 5.13). In the real world, things are less simple. Using the example of daily Central England Temperatures since 1772, we have already noted that there has been a marked winter warming.[28] In Fig. 5.14 the distribution of values for January for the fifty years 1772 to 1821 is compared with those for 1946 to 1995. For both periods it is distinctly skewed. But, although the latter period was nearly 1.4 °C warmer than the earlier one, the range of extremes is virtually unchanged. The real differences are in the central regions where, on the smoothed curves, the median has shifted 3.5 °C reflecting that there are substantially more cold days in the first period and more mild days in the latter. Analysis of December temperatures shows the same behaviour, but February has shown no appreciable warming and the distribution during the two periods is effectively identical.

One interpretation of these figures is that, in mid-latitudes of the northern hemisphere at least, global warming is associated with shifts in the frequency of different weather regimes rather than any more fundamental change in the nature of these regimes. On the face of it, the observations confirm a lasting shift to more westerly weather and away from blocked

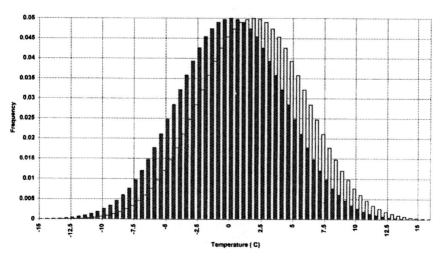

Figure 5.13. A theoretical representation of how the distribution of daily temperatures for a given month at a given site might shift with a warming of 2 °C, assuming that the distribution was 'normal'.

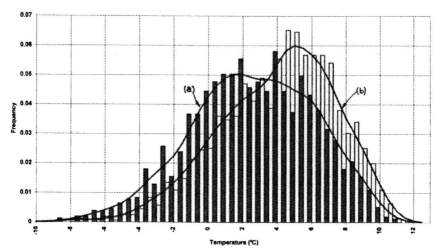

Figure 5.14. The distribution of Central England (daily) temperatures in January for (a) 1772 to 1821 and (b) 1946 to 1995, showing a shift in average temperature of around 1.5 °C.

easterly conditions. If this is the consequence of human activities warming the climate then it might be as well to plan on this trend continuing. As we will see in Chapter 6, however, computer models have not demonstrated yet that they can reproduce the statistics of quasi-stationary weather regimes. Moreover, the non-linear response of the climate to change could have the perverse result of a warmer world featuring more blocking in winter and hence colder winters for parts of the northern hemisphere.[29] But, for the moment, this is pure speculation. In the meantime, it is best to assume that, while the incidence of different regimes is liable to change, past experience remains relevant to future plans, although the balance of events seems likely to continue in the direction of warmer conditions.

5.8 Challenges for climatologists and climate modellers

The brief résumé of the changes that have occurred in the climate in the last millennium or so and its statistical characteristics, provides a background for considering what is needed to improve predictions of future climatic change. The questions that remain unanswered about nature and causes of past climatic variability are a major challenge to climatologists. If we could obtain better measurements of what happened in the past then we could get a more reliable handle on how much of the variability is the natural product of the complex coupling between the atmosphere and the oceans and how much is due to external agencies. In respect of the latter, two particular influences are of great interest – major volcanic eruptions and solar activity. The sooner we have improved insights into these questions of natural climatic variability the sooner we can be confident that models of the global climate are capable of providing a useful representation of the impact of human activities (see Section 6.2).

The question of the natural variability of the climate hinges principally on two things. First, there is the question of what controls the incidence of weather regimes, which has been discussed in the preceding section. The second is improving our understanding of the longer term variations in the oceans. These two may be connected, but here we will concentrate on the latter. The interannual fluctuations of the equatorial Pacific – the ENSO – have already been discussed in respect of droughts throughout the tropics in Chapter 4, while the possibility of more radical variations in ocean heat transport was touched on earlier in this chapter. At present we know far too little about what controls sea surface temperatures on timescales of decades and longer, and how the atmosphere combines with the oceans both to help drive these changes and to react to them. We need to know whether quasi-periodic interactions could have combined to produce the colder conditions of the 1590s, 1690s and 1810s, with their different mix of weather regimes.

The possibility of the climate system combining to produce sudden shifts is of particular relevance now. This effect may have been at work in the 1980s and 1990s to produce the exceptional warmth of the last 15 years or so. Many climatologists have commented on the sharp switch in circulation patterns that occurred in 1976. The period since then has been marked by far more frequent ENSO conditions and by much stronger westerly circulation during the northern hemisphere winter. Alternatively, it may be that human activities have warmed the atmosphere in such a way as to switch the equatorial Pacific into a quasi-permanent ENSO state. The capacity of current climate models to handle the behaviour of the ENSO and predict future developments in the Pacific has to be a crucial test of their utility in generating economically useful forecasts (see Section 6.2).

When it comes to shifts in larger scale heat transport of the oceans, we move into even more speculative areas. The possibility that the North Atlantic could be part of bigger changes is the starting point. It may be linked to the existence of the Great Ocean Conveyor (GOC) (Fig. 5.15) and the possibility that it can operate in a number of different states. Thanks to the early work of Wallace Broecker at the Lamont-Doherty Earth Observatory of Columbia University in New York, this has become a hot topic in the understanding of climatic change.[30] Because GOC plays such an important role in the climate of the North Atlantic it is essential that we discover more about its behaviour. At present we know that the circulation in the twentieth century carries huge amounts of heat up towards Greenland. This warm water gives up its heat to the arctic air through evaporation and by direct warming of the colder atmosphere. This causes its temperature to fall, and its salinity to rise, thereby increasing its density. This denser water sinks and forms 'deep water' which flows back all the way to Antarctica. Here it is warmer and less dense than the frigid surface water so it rises, and becomes part of a strong vertical circulation process. Descending cold water from around Antarctica flows

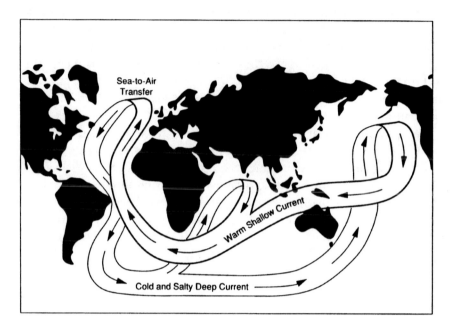

Figure 5.15. The Great Ocean Conveyor Belt – a schematic diagram depicting
global thermohaline circulation. (From Trenberth, 1992, Figure 17.12.)

northwards into the Pacific and Indian Oceans where there is no
descending cold water. In the Atlantic this countercurrent is swallowed
up by the much stronger southward flow.

Recent modelling work[31] has suggested that the amount of deep water
formation and where it is formed is extremely sensitive to the amount of
fresh water entering the North Atlantic. This can vary as a function of
run-off from the continents, the amount of ice calved from the Greenland
ice sheet, and rainfall from weather systems. Small changes in the total
input may be able to trigger sudden switches to alternative conveyor belt
patterns which mean that less heat is transported to the most northern
parts of the Atlantic Ocean. This could reduce the temperature of the
waters around southern Greenland and Iceland by 5 °C or more, which
would have a drastic impact on the climate of Europe and completely alter
the weather patterns of the northern hemisphere. It would also make the
experience of past cold winters in Europe of far greater relevance than is
currently assumed.

The possibility that these processes could explain the sudden changes which occurred during the last ice age (see Section 5.2), and possibly during the previous interglacial, makes these modelling results a chilling reminder of the scale of our ignorance of what drives long-term changes in the circulation of the oceans. It can be argued that what drove the changes during the last ice ages was sudden massive outbursts of icebergs from the ice sheets covering Labrador and Greenland, which created the Heinrich layers that are observed in ocean sediment cores from the North Atlantic (see Fig. 5.2).[32] So it may be that the existence of major ice sheets around the North Atlantic is a necessary condition for the erratic behaviour of the GOC. But the fact that we may be triggering a change in the climate which exceeds anything seen in the last 10 000 years means that there is a desperate need to find out more about what governs the global transport of heat by the oceans, and how it might be influenced by continued global warming.

Even if sudden switches in the GOC are not to be part of the immediate future, this circulation provides an explanation for climatic fluctuations on the scale of decades and centuries. It is estimated that the time taken for water to flow round this global loop is of the order of 500 to 2000 years; it can therefore act as a huge climatic flywheel. Changes in the amount of deep cold water created in the North Atlantic centuries ago and its temperature can be stored in its memory to re-emerge in the future to act as a driving force for natural variability of the climate. So finding out more about how the GOC functions and what is locked up in its memory is high on the research agenda.

By comparison with the depths of our ignorance about the oceans, in one respect our understanding of the climatic impact of external agencies is much improved of late. This is because of the massive eruption of Mount Pinatubo in the Philippines in June 1991. This event, which injected some 20 million tonnes of sulphur compounds into the stratosphere, was by far the biggest eruption this century. As such it provided an excellent opportunity for testing theories about the impact of volcanoes on the climate. This issue was given additional flavour at the time by James Hansen of the Goddard Institute of Space Studies in New York, who predicted on the basis of modelling work that within a year Pinatubo would cool the global climate by 0.5 °C and then the temperature would return to normal within three years or so. Both satellite and surface-based observations (Fig. 5.16) confirmed how accurate this prediction was in the size and duration of this eruption.

This result has a couple of interesting implications for atmospheric

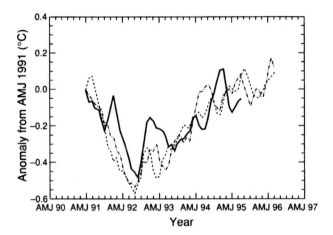

Figure 5.16. The observed cooling at the Earth's surface attributed to the eruption of Mount Pinatubo in June 1991 (solid line) together with predictions of cooling which was expected to occur on the basis of the simulations by computer models in 1992. (Hansen *et al.*, 1992).

sciences. First, it confirms volcanic eruptions do play a significant, if short-lived, part in climatic change, and verifies that the massive eruption of Tambora in Indonesia, which injected five to ten times more material into the stratosphere than Pinatubo, probably played a major part in the extreme weather in 1816 (see Section 2.4). It suggests that, in the right circumstances, major volcanoes can exert longer term influences on the climate. An extreme example of this process is the largest eruption in the last million years – Toba in Sumatra, 73 000 years ago, which was at least five to ten times the size of Tambora – coincided with one of the most rapid cooling stages during the onset of the last ice age. Modelling work suggests that the cooling was sufficient to produce perennial snow cover for a number of years over northern Canada.[33] At a time when the Earth was slipping into an ice age, the additional cooling resulting from snow reflecting sunlight back into space throughout the summer may well have been enough to tip the balance towards the onset of glacial conditions. So it may be possible for major eruptions to trigger more lasting changes in the climate, if the timing is right.

The second point is how well computer models of the global climate (see Section 6.2) performed in predicting the impact of Pinatubo. If they are capable of handling the type of perturbation caused by volcanoes, it increases our confidence in their predictions of the impact of human

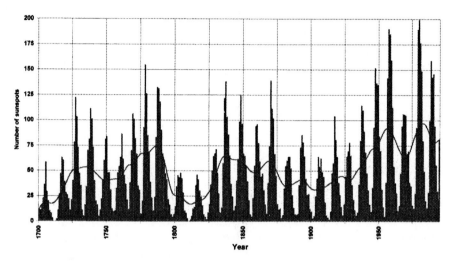

Figure 5.17. The variation of the average number of sunspots each year between 1700 and 1995, together with smoothed data showing longer term fluctuations.

activities on the climate. It also implies that if the forecasts of future warming prove to be correct, volcanoes are unlikely to exert a lasting impact on future trends. Rather, they will provide short-term and possibly spectacular setbacks, but will not prove to alter the course of future climatic change.

The other possible external agent of climatic change which exerts an undying fascination is solar activity and its most visible manifestation – sunspots. Ever since the astronomer William Herschel first proposed, in the early nineteenth century, that changes in the Sun's output could influence the weather, and then in 1843 Heinrich Schwabe discovered that the number of sunspots varied in a regular, predictable way, there has been a continual effort to demonstrate that the cyclic behaviour of solar activity is reflected in variations in the weather. There is not space enough here to go over the history and current status of all the statistics that have been collected to demonstrate that a connection exists or the arguments that have been used to demolish the claimed significance of these connections.[33] Suffice it to say, there remains a striking parallelism between observed global warming in the twentieth century and solar activity and also in respect of the cool period in the 1810s (Fig. 5.17).[34] Furthermore, there is intriguing evidence of the cooler episode in the late seventeenth century coinciding with solar activity being quiescent with a virtual absence of sunspots (the Maunder minimum).[35] So the current state of the debate on whether observed fluctuations in solar activity has an impact on the climate has to be reviewed briefly here, not least because until this issue is resolved, it will continue to bedevil discussion on the extent to which human activities are responsible for current global warming.

The focal point for the current debate is whether the well-established but small changes in the output of the Sun during the 11-year sunspot cycle are capable of having an appreciable impact on the climate. At the simplest level the argument is clear cut. The amplitude of observed changes during the 11-year cycle amount to less than 0.1 per cent of the total output of the Sun. In crude energy terms this is far too small to account for the observed changes in global temperature over the last 100 years or so. Either the changes in the Sun's output are not responsible for a significant part of the observed changes or more subtle physical processes are at work to amplify the effects of the Sun. One such possibility is related to the fact that much of the Sun's variability is concentrated in the ultraviolet (UV) spectral region. Although only one per cent of the Sun's energy is emitted in the 200 to 300 nanometre range, some 20 per cent of the observed change in solar output is concentrated in this range. This energetic radiation is absorbed high in the atmosphere by oxygen and ozone and is also instrumental in producing ozone at these levels.

Joanna Haigh, of the Space and Atmospheric Physics Group at Imperial College, London, has used a computer model of the atmosphere to calculate how the amount of solar radiation entering the lower atmosphere varies with solar activity and how this might affect the circulation of the atmosphere.[36] Changes in the ozone concentrations in the stratosphere caused by the changing UV flux may lead to an amplification process with the amount of solar energy reaching the lower atmosphere in middle and high latitudes in winter being appreciably reduced when solar activity is high. Computer simulations suggest that the effect of these changes on global circulation could be to alter the position of the upper atmosphere jet streams and hence the mid-latitude winter storm tracks. This may help to explain why analysis of storms north of 50°N in the North Atlantic between 1921 and 1976 showed that they were on average 2.5 degrees of latitude further south at the sunspot maxima than at the sunspot minima[37].

Changes in the amount of UV reaching the lower atmosphere may also exert photochemical effects, which have far-reaching ramifications. Modelling work by Ralf Toumi in the Department of Physics at Imperial College, and Slimane Bekki and Kathy Law at the University of Cambridge predicts chain reactions resulting from reductions in ozone levels in the stratosphere.[38] These changes increase the number of cloud condensation nuclei, which can increase cloudiness and hence lead to cooling of the climate. These studies conclude that perturbations in the amount of ozone in the upper atmosphere will have a disproportionate effect on the climate.

There are other proposals for effects relating to the magnetic fields associated with solar activity, which may also modulate the Earth's climate.[39] While plausible, these alternative propositions do not alter the fundamental issue – until the part that solar activity has played in the natural variability of the climate is clearly identified, there will be those who argue that it is a significant contributor to recent changes in global temperature. This uncertainty, however much the consensus view is that human activities are the principal factor in events, is bound to make it more difficult to reach difficult decisions on action to prevent future climatic change.

5.9 Determinism

Finally, before getting down to the serious business of discussing what, if anything, we can forecast about the weather and its economic and political consequences, we lay to rest the ghost of determinism. Having defined the challenges facing the climatological community, it is important to be clear how much progress is feasible. There is still a school of thought – buoyed up by claims of huge advances in parallel computing, offering speeds measured in umpteen petaflops per second within a few decades – that is unabashed by the insights about the unpredictable nature of nonlinear systems provided by Chaos Theory. The hope is that with improved measurements and faster computers it will be possible to feed more precise details of the state of the atmosphere into the models and produce ever more accurate forecasts. While this no longer goes as far as the Laplacian view that if we had complete knowledge of the position of every particle in the universe and its motion, using the Laws of Nature, we could forecast the future for ever and ever, it still assumes that progress will in some way be proportional to increases in computing power. It is instructive to consider where this thinking leads us.

Long before we get to the level of worrying about the fabled butterfly of Chaos Theory whose flapping wings may eventually influence the weather on the other side of the globe,[40] we will have ventured further afield. Whether or not the effects of sunspots are amplified in the atmosphere to produce a significant impact on the weather, there is no doubt that the amount of energy emitted by the Sun during the 11-year cycle in solar activity varies by about 0.1 per cent. While this may seem small, in a deterministic world it is massive, because to make accurate predictions we must have precise knowledge of all factors that could influence the predictability of the weather events in the longer term. So, in the search for greater precision, our forecasts would need to include predictions of solar activity on all timescales. Beyond this we will need to include the tidal effects of the Sun, the Moon and of other planets in the solar system, notably the pull of the giant planets on both the Earth and the Sun. It follows that in due course, along with the predictions of solar weather, our model would need to include predictions of the behaviour of the giant red spot on Jupiter. Inexorably this leads to not only knowing what butterflies on Earth are up to, but also what is going on throughout the rest of the Universe.

This may all seem a mite whimsical, but it contains an important point. We will never be able to know enough about all the factors that matter in predicting the climate. To use another example, in a ground-breaking paper on chaos, which appeared in *Scientific American* in December 1986, postulated an idealised experiment of a billiard table with perfectly elastic balls and cushions and no resistance to motion.[41] If one ball was set in motion, and the intention was to predict the precise movement of the balls, moving at normal speeds, after a minute it would be necessary to include the gravitational effect of an electron on the far side of the Universe.

This extraordinary conclusion arises from the extreme non-linearity of the process of billiard balls making elastic collisions. Consider an even simpler experiment with ten balls in a long line on a special elongated table. Then ask what accuracy is needed to ensure that on hitting the first ball it will collide with the second, which will then strike the third, and so on until the ninth strikes the tenth? The enormity of this task soon becomes evident. As can be seen in Fig. 5.18, a small error in the direction of the first ball (α radians) is magnified at the collision with the second ball by the ratio of the diameter of the balls (d) to their distance apart (D) (i.e. D/d). If the balls are well separated so that D is $100d$ (in the case of billiard balls about five metres apart) then the error grows one hundredfold at each collision. This means that merely to ensure that the ninth

(a)

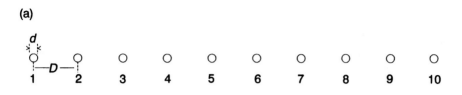

(b)

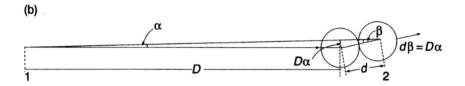

Figure 5.18. The challenge of striking the first of ten billiard balls so as to ensure that they all hit the next one in the line can be seen in this pair of diagrams where (a) shows ten identical balls (diameter d) in a straight line equally spaced D apart, with D being much larger than d, then (b) shows what happens when ball 1 is hit with at a tiny angle (α radians) to the line joining the centres of all ten balls. By the time it reaches the second ball it is αD off course. So when it strikes ball 2 it sends it at an angle β to the intended course where $\beta = (D/d)\alpha$. This means the initial error α is amplified by D/d at each collision until one ball misses the next one. For ball 9 to hit ball 10, α must be less than $(d/D)^9$ radians.

ball hits the tenth the angle α has to be less than 10^{-18} radians. This is a stupefyingly small angle.

To explore what this accuracy means, it helps to consider what factors will come into play along the way to achieving success. For instance, failure to take account of the forces due to the Earth's rotation will introduce an error of around 1 part in 10^4 and so mean that even getting the second ball to hit the third we would have to consider the latitude at which the experiment is conducted. As for the sphericity of the balls, this is probably no more accurate, and even with balls whose sphericity could be defined to a single atom's width, we are still only talking of around one part in 10^8, by which time air currents, the local gravitational effects and the radiational pressure of the lighting would start to come into play. To achieve the additional ten orders of magnitude in accuracy will require taking account of the rest of the solar system. And yet we are still only considering the elemental task of getting ten billiard balls to knock into one another when set up in a straight line. By comparison the task of predicting precisely what the atmosphere is going to do for a few weeks ahead adds many, many more orders of magnitude of precision to our starting conditions.

But enough of these flights of fancy. What matters here is that in facing up to the challenges of deciding which factors matter in predicting the weather and climatic change, and their economic and political conse-quences, we have to accept we are operating in an uncertain world. As a consequence, errors in our initial conditions will grow in any prediction we make. In the case of the weather, if we represent the climate system in a suitably large lump we can treat it in an effectively deterministic manner a few days ahead (see Section 6.1). Beyond this, the growth in errors due to the non-linearity of the atmosphere eventually swamps the forecasts; any chance of accurately foretelling how the day-to-day weather will develop is impossible and we enter the realms of statistics. Here we can still make predictions providing we know enough about the properties of the system we are simulating. It is like throwing a perfectly balanced die 600 times. We can predict that each number from one to six will come up about 100 times. But we can never predict in advance the precise sequence of numbers which will occur with each successive throw. In the same way, if we know enough about what controls the statistics of the global climate then we can make useful predictions about how these will change if perturbed by human activities. In so doing, we will always be talking in terms of the balance of probabilities.

In working in this statistical world, two criteria must be met. First, we have to have a clear statement of what the forecasts are capable of doing. This boils down to how any prediction presents the statistical uncertain-ties which are inherent in its production. Without some indication of the 'skill' of the forecasts,[42] there is a danger of their predictions being grossly misleading. To embark safely on modelling and forecasting work, we must keep the fundamentally indeterminate nature of the systems under exam-ination in the forefront on our minds.

The second issue relates to the challenges identified earlier in this chap-ter. It is whether, in deciding which factors matter and which can be ignored in order to produce manageable models of the climate, there is a bias towards certain outcomes – in effect, whether the computational dice are loaded. As in the real world, where loaded dice will eventually show up in observation of the distribution of outcomes, any bias in the com-

puter models will similarly become clear over time. The trouble is that
we want to use the models now and any checks of their ability to reproduce the statistical nature of climate change will take a very long time. So
all we can do is to extend our requirement of statements of 'skill' to cover
the question of what has and has not been included in any prediction.
This places an additional onus on forecasters to spell out the limitations
of their models. It also provides us with the basic framework for exploring
the world of climatic and economic models.

5.10 Notes

1 Imbrie & Imbrie (1980).
2 There has been an avalanche of papers on various aspects of the evidence
   of rapid climatic change in ice cores and other proxy records in recent
   years. The current status of this work is reviewed in Section 3.6.3 of
   Chapter 3 of IPCC (1995); and other important papers include Alley *et al.*
   (1993), GRIP (1993), Grootes *et al.* (1993) and Taylor *et al.* (1993).
3 GRIP (1993), and Grootes *et al.* (1993).
4 GRIP (1993).
5 Grootes *et al.* (1993).
6 Taylor *et al* (1993).
7 Broecker (1994).
8 Lorenz (1993).
9 Lamb (1995), p. 175.
10 See Chapter 19 by Briffa & Schweingruber in Bradley & Jones (1995).
11 Briffa *et al.* (1995).
12 Lamb (1995), p. 195.
13 IPCC (1992), p. 141.
14 See Chapter 6 by Pfister in Bradley & Jones (1995).
15 Le Roy Ladurie & Baulant (1980).
16 Van den Dool, Krijnen & Schuurmans (1978).
17 Manley (1974).
18 Engelen & Nellestijn (1995).
19 IPCC (1992), p. 141.
20 Bradley & Jones (1995), p. 660.
21 Folland & Parker (1995).
22 Spencer & Christy (1990).

23 Hansen *et al.* (1995).

24 Karl *et al.* (1996).

25 IPCC (1995), p. 168.

26 Burroughs (1994), Chapter 8.

27 Lamb (1972), p. 243.

28 Parker *et al.* (1992).

29 Palmer (1993).

30 Broecker (1995a).

31 Weaver & Hughes (1994).

32 Dansgaard *et al.* (1993).

33 Rampino & Self (1992).

34 Burroughs (1994), Chapter 6.

35 Eddy (1976).

36 Haigh (1996).

37 Tinsley (1988).

38 Toumi, Bekki & Law (1995).

39 Markson (1978).

40 The origin of this icon of Chaos Theory is discussed in Lorenz (1993), p. 14. This book also provides an accessible analysis of the contribution meteorology made to understanding the impossibility of finding deterministic solutions to the behaviour of non-linear systems, and the part Edward Lorenz played in establishing this fact.

41 Crutchfield *et al.* (1986).

42 Tennekes (1992).

6

Models of the climate and the economy

'Write that down,' the King said to the jury, and the
jury eagerly wrote down all three dates on their slates,
and then added them up, and reduced the answer to
shillings and pence.
Alice's Adventures in Wonderland, Chapter 11

So far we have explored a wide variety of examples of the interaction between extreme weather events, both fleeting and long drawn out, and economic and social developments. While these clearly show that the weather can wreak havoc on all timescales, the nature of this impact varies profoundly from event to event. This means that any attempt to predict the economic consequences of future climatic change faces major challenges. Any forecast must identify with considerable precision not only what type of developments are likely to become more prevalent, but also how out of the ordinary they are likely to be, given the non-linear relationship between such events and the damage they cause. Even if the climate forecasts can rise to this challenge, there is then the next level of complexity of how economic and political structures should respond to sustained abnormal weather.

It would be easy to throw in the towel at this stage. Clearly the complexity of the impacts on society and the responses they evoke are far beyond the current capacity of forecasting systems. But, as noted in Chapter 1, we cannot adopt the defeatist attitude, however realistic it may seem. The fact of the matter is that those who make political decisions have to demonstrate they have made effective use of expert or scientific opinion and the available forecasting tools, whatever their limitations. So we must examine these instruments and consider their predictive value. This is not a beauty contest. The aim is not to show that one or other set of models does a better job of tackling its area of interest, but to explore each area's strengths and weaknesses. Only then can we reach conclusions as to how planners and policy-makers should be using the combined product of this work.

In making this assessment there are some general observations on the nature of models to make first. The most important point is that they provide a means to test hypotheses and to order thinking. As such they can range from on the one hand very simple systems to explore the potential consequences of making changes in specific parts of economic or physical structures to, on the other hand, three-dimensional, multivariate, time-dependent models which use vast amounts of time on the most massive supercomputers. It is the latter that are of particular interest here, but the general point remains that at whatever the level they operate, they constitute a numerical experiment to explore a variety of 'what if?' questions. So, at the first stage, our interest is not with whether the models are 'correct' but with what they tell us about how the global climate and economy function and their capacity to spring surprises on us.

Beyond the exploratory phase, it is natural to proceed to the verification phase of checking how the models perform in practice. But, in talking about verification the best we can hope for is that the results of the models ring true. It is not possible to encapsulate the real world in a model. By making a series of approximations and simplifying assumptions, what we have is a tool that can provide insight into how the world goes round, but can track reality in only a limited way. Although we can set measures of performance in terms of the match between certain parameters in the model and in the real world and how well forecasts predict the future, this cannot be extended too far. Improvements in, say, models for forecasting the weather a few days ahead can be measured. As these exercises are extended further into the future, however, the problem of defining which factors matter most comes to dominate. It is not possible to adopt an even remotely deterministic view about predicting the climatic change or economic change (see Section 5.9). The best we can hope for is that the models have succeeded in capturing the most essential features in what is driving the world around us. If this is the case then there is a reasonable chance that the factors over which we can exercise some policy control will influence the future in roughly the manner predicted. What is certain is that it is hard to imagine the conduct of economic policy without predictions. Either explicitly or implicitly the conduct of any policy implies a view about the future.

6.1 Numerical weather forecasting

Computer models of the global climate have been a major factor in the development of meteorology since the 1950s. Indeed the insatiable appetite for greater computer power to handle bigger and better models has been one of the principal pressures for the development of massive supercomputers. The global models that have issued from these computers have played a dominant role in two particular areas. First, they have become central to standard weather forecasting. Secondly, they are the foundation for studying the impact of human activities on the global climate. Their performance in both these areas provides insight into their strengths and weaknesses.

Computer models treat weather forecasting as a problem in mathematical physics. They contain a very large and complex array of equations based on the physical and dynamical laws which govern the birth, growth, decay and movement of weather systems. They incorporate the principles of conservation of momentum, mass, energy and water in all its phases; the Newtonian equations of motion applied to air masses; the laws of thermodynamics and radiation for incoming solar energy and outgoing heat radiation; and the equations of state of atmospheric gases. Parameters which are specified in advance include the sizes, rotation, geography and topography of the Earth, the incoming solar radiation and its diurnal and seasonal variations, the radiative and heat conductive properties of the land surface according to the nature of the soil, vegetation and snow and ice cover, and the surface temperature of the oceans.[1]

The physical state of the atmosphere is updated continually drawing on data from observations from around the world via surface land stations, ships, and buoys, and from the upper atmosphere using instruments on aircraft, balloons and satellites. Just assimilating this data involves four trillion calculations per forecast. The model atmosphere is divided into thirty layers between the ground and an altitude of around 30 km, and, in the most advanced models, each level is divided up into a network of points about 50 km – apart some four million points in all. Each of these points is assigned new values of temperature, pressure, wind and humidity with each new run of the model and the governing differential equations are integrated in fifteen-minute steps at each point to provide forecast values up to ten days ahead. Each new set of forecasts involves about twenty trillion calculations. The output is hundreds of forecast charts of pressure, temperature, wind, humidity, vertical motion and rainfall which are used to provide a variety of forecasts for many different customers and the general public.

These immense models have made substantial progress in producing useful forecasts of the instantaneous weather up to about a week ahead. The big players in this game are the German and US weather services, the UK Meteorological Office and the European Centre for Medium Range Weather Forecasts (ECMWF). The products of these agencies are widely used by national forecasting agencies, the media, and other public and private bodies. In terms of the shorter term forecasts tailored for regional requirements the national services usually produce the best efforts. When it comes to forecasts three to five days ahead, which involve analysis of global weather patterns, it is reasonable to compare the predictions of different agencies. How far this goes can be gauged from a forecast I heard on a Boston radio station in early March 1996. The forecaster noted that the US Weather Service was predicting that a storm at the end of the week would produce rain, while the ECMWF was going for snow and ice. At the end of what had been a record snowy winter with frequent coastal storms, the forecaster plumped for the ECMWF version on the basis of earlier successes: he was right – Boston got another 25 cm (10 inches) of snow. This result is not a criticism of the US forecast, but a reflection of how the ECMWF service operates. Because it produces forecasts two to 10 days ahead, it does not have to rush to get short-term predictions out, and has more time to process all the available data before running its model. This means it forms a more accurate picture of current conditions, which pays dividends a few days ahead, but at the price of taking longer to produce its forecast.

The importance of forming as accurate a picture of current weather patterns as possible is central to better forecasts. There are, however, inherent limitations in the predictability of atmospheric behaviour.[2] These properties have been examined by comparing two or more outputs of the models with virtually the same initial conditions, and observing the growth in the difference between the solutions. These differences double every two days or so. This implies that the limit to producing useful forecasts of the instantaneous weather is 10 to 14 days ahead, even with greatly improved observations of the initial state of the atmosphere.[3] Clearly, for the purposes of predicting climatic change this is a barrier which must be circumvented.

The possibility of handling the unpredictable behaviour of the atmosphere leads into two aspects of the statistical nature of longer term forecasts. First, on the immediate question of getting the best out of weather forecasting efforts, the speed with which different predictions with similar initial conditions diverge over the forecast period provides an important insight into the predictability of current weather patterns: in less predictable situations the differences develop rapidly, while with more predictable conditions they grow more slowly. Secondly, there is the related issue of whether or not the weather is stuck in a stable regime (see Section 5.7). Each winter in the mid-latitudes and the northern hemisphere, three-quarters of the time the atmosphere is in one of four or five of these patterns. Once established, a given regime may persist for several days or longer. During such a quasi-stationary situation the weather behaves in a more predictable manner. But when the regime breaks down, the change is rapid and often unexpected.[4]

Relative stability punctuated by sudden and less predictable changes is the challenge confronting modellers. It has profound implications for both day-to-day forecasting and predicting climatic change. In the case of weather forecasting, it means the quality of the forecasts depends on the sensitivity to the uncertainties in measurement of the initial state of the atmosphere. This sensitivity is a property of whether the atmosphere is in transition between quasi-stationary states or stuck in one regime or another. One way of finding out whether a switch in regimes is likely to occur during the forecast period is to see how the predictions respond to slightly different starting conditions. If, with a subtle range of starting conditions, an ensemble of forecasts looks remarkably similar up to, say, 10 days ahead, then there is a good chance they are on the right track. If, however, the different forecasts diverge significantly, then clearly the atmosphere is in a less predictable mood. This ensemble forecasting technique is now an integral part of the weather forecaster's armoury – the ECMWF now runs 50 versions of its standard 10-day forecast. As a result of this progress, useful predictions can now be produced up to around six days ahead.

6.2 Global climate models

When it comes to predicting the possible climatic consequences of long-term perturbations of atmosphere and the oceans by either human activity or natural agencies, slightly different approaches need to be adopted.

Interest switches to the broad sweep of weather patterns and the statistical aspects of weather regimes in special global climate models, which are usually called General Circulation Models (GCMs). These models use the same set of mathematics and physics involved in the weather forecasting work. But in computing the consequences of changing certain aspects of the climate system the models have to simulate much longer periods. This has two principal consequences. First, it requires the GCMs to simulate the behaviour of the oceans and how they interact with the atmosphere. Some of these processes are included in weather forecasting, but predictions up to 10 days ahead can use current sea surface temperatures and assume they effectively remain constant. Simulating the global climate for centuries ahead requires the oceans to be represented with the same realism as the atmosphere.

The second issue is simply that time is money. Supercomputers cost a great deal to run, and so, in the interests of keeping the computation time manageable and within budget, the models have to cut down on the detail by using a larger grid spacing and by lengthening the time intervals between successive computations. These changes are linked by the way the models operate and so reduction in resolution, both spatial and temporal, is at the cost of simplifying the representation of the climate.

Typical GCMs for climate prediction work at the end of the 1980s had a grid spacing of 4.5 degrees of latitude by 7.5 degrees of longitude (500 by 600 km at 40° N or S). With nine vertical layers this provided 17 200 grid point to represent the atmosphere very much less detail than the weather forecasts described earlier (Fig. 6.1). But with the computations conducted for 30 minute intervals in the model, it still took some 10 hours on the largest supercomputer at the time (Cray X-MP) to simulate one year of global climate.[5] Given that many modelling exercises required decades of simulation to get a decent feel of how perturbations influenced the climate, and in some cases hundreds of years to examine the variability of the system, it can be seen that computing time and costs have been a limiting factor in this work.

Recent advances in the power and speed of the largest computers have permitted the development of more sophisticated GCMs. Part of this progress has been used in producing higher resolution representations of the atmosphere. But most effort has been devoted to incorporating the oceans into the models. Because 'weather' in the oceans involves much sharper, smaller features (i.e. currents and eddies), ideally the models should be more detailed. The highest resolution research models of the

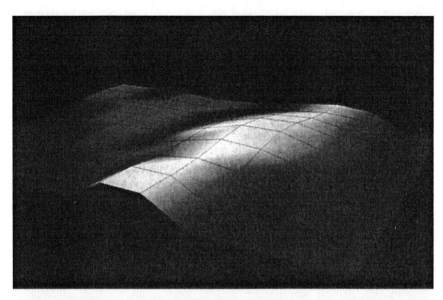

Figure 6.1. A schematic depiction of the topography over North America as represented in most coarse-resolution (480 km grid) atmospheric general circulation models for climate simulation (top) and in high resolution (60 km grid) global numerical weather forecasting models (bottom), showing how much detail is lost in performing the much longer simulations of the global climate. (Reproduced by permission of Thomas Bettge, NCAR.)

oceans have up to 60 levels and resolutions as fine as one-sixth of a degree of latitude and longitude.[6] While these can produce realistic representations of many of the oceans' dynamics, they are far too expensive and slow to incorporate current climate studies. So coupled atmosphere–ocean GCMs have comparable resolution for both components. But, this compromise does risk throwing the baby out with the bathwater because it requires a physical fudge to handle the finer details of currents and eddies in lengthy simulations of the climate.

The essence of this compromise is how the models deal with fluxes of heat, momentum and fresh water between the oceans and the atmosphere. These will vary dramatically throughout the year and from place to place. For instance, the oceans will provide large amounts of heat to atmosphere in winter and absorb it in summer. These exchanges involve very large numbers, and inevitably small errors build up over time, leading to the models drifting away from climatic normality.[7] To prevent this happening some models make pragmatic adjustments, known as 'flux adjustments' to keep them on the climatic straight and narrow.

Improved ocean–atmosphere coupled GCMs is a world-wide pursuit. The 1995 IPCC Report lists 16 models being developed in Europe, North America, Australia, Japan and China. Among the most advanced examples is the one used by the UK Meteorological Office Hadley Centre at Bracknell, which consists of a coupled ocean–atmosphere model with a resolution of 2.5° latitude by 3.75° longitude, the ocean represented by 20 layers and 19 layers for the atmosphere.[8] The treatment of the dynamics of the atmosphere is effectively the same as in the weather forecasting models. It tackles the flux adjustment issue by making calibrated seasonal flux adjustments to seasonal SSTs and ocean salinities to bring about a faithful representation of the present mean global climate. Also there is a simple treatment of sea-ice drift to remove the need for making regular flux adjustments for this important climatic parameter.

The alternative to fudging the numbers is to let the model wander off on its own into what may be unreal vapourings. This may seem a drastic course, but it does have the merit of retaining the best possible models on purely physical grounds. At the moment, there is no simple answer. Only when the models can provide sufficient resolution to handle the fluxes associated with, say, the major ocean currents will the dilemma be resolved. Until then you pay your money and take your choice – either you can have a model that looks realistic but, which may have suppressed some important aspect of climatic variability, or one that you believe is a physically sound construction but which may live out much of its existence in the realms of fantasy.

Other challenges confronting the climate modellers relate to the more accurate representation of the physics of certain aspects of the atmosphere, oceans and the Earth's surface. Most important of all these problems is the handling of clouds. The reasons are obvious. The combination of their transient behaviour, wide variety of forms, and the dominant role they play in the radiation balance of the Earth means that they need to be accurately represented in any model of the climate. Furthermore, the fact that on many occasions they come in sizes far smaller than the grid spacing in the models means they also have to be subjected to some form of averaging analysis to produce a collective representation – usually termed parameterisation. This technique is similar to the empirical adjustments made in the slowly varying components of the climate. But, in the case of clouds, we are concerned with a rapidly varying part of atmospheric dynamics which has to be accurately portrayed in all modelling work.

The scale of the challenge posed by clouds for atmospheric physicists can be gauged by the recent debate about the radiative properties of clouds.[9] Clouds exert two fundamental and competing influences on the climate. First they scatter and reflect sunlight. Because some solar radiation is reflected out into space, clouds have a substantial cooling effect. This depends principally on the number of ice crystals or droplets present and their size. Second, they are strong absorbers and re-emitter of the heat radiation from the Earth, which has a warming effect. The amount of warming is a function of the temperature of the cloud tops. Low decks of stratus, whose temperature is little different from the surface, have only a small warming effect, but are efficient reflectors of sunshine. Conversely, high thin cirrus reflects a modest amount of sunlight, but because its low temperature reduces the amount of heat radiated to space appreciably, it has a net warming effect. In between, the gamut of cloud types from towering cumulonimbus to puffy little fair-weather cumulus exert an intermediate impact. While the overall effect of an increase in cloudiness will have a cooling effect, much depends on the precise nature of the clouds created.

Recent measurements have demonstrated how complex these processes may be.[10] They suggest that clouds absorb far more sunlight than previously assumed. By making simultaneous measurements in aircraft flying both above and below clouds, together with ground-based and satellite observations, researchers concluded that clouds absorb four times more energy than expected. This posed huge problems for climate modellers. It would require a significantly different set of atmospheric dynamics, especially in the tropics, as less energy would reach the surface and much more would be injected at higher levels. This would reduce convection and rainfall, and with this the efficiency of the tropical boiler which drives much of the global climate system.

When these new figures were put into the GCMs they had a mixed impact. In the case of the model designed by the National Center for Atmospheric Research (NCAR) at Boulder, Colorado, USA, the results were positive.[11] Because this model tended to have a rather wet, stormy climate, absorbent clouds toned down the weather and made it look more like the real world. With the Hadley Centre model it turned down the wick too much, so that what had looked a pretty good representation of the climate became too dry.[12] The differential response of various models is a measure of their limitations, and the question of the physical properties of clouds is currently the subject of intense scientific investigation. A similar picture emerged from a study of the sensitivity of 17 models to changes in snow cover.[13] Because snow is an efficient reflector of sunlight, and global warming would reduce the amount of snow and hence increase the amount of solar radiation absorbed at the surface, it has been widely assumed that this positive feedback would accelerate the warming process. While the majority of the GCMs supported this hypothesis, a few had a negative feedback principally because increased cloudiness at high latitudes tended to cancel out the reduction in snow cover.

In spite of these limitations, GCMs are the only way to calculate how human activities may alter the climate and to explore competing theories about how severe the consequences may be. Comparisons of the performance of various GCMs shows they provide a realistic portrayal of the phase and amplitude of the seasonal march of the broad features of pressure, temperature, and circulation.[14] The largest discrepancies occur in sea-level pressure and surface air temperature at high latitudes and in precipitation in the tropics. The scale of these discrepancies is large compared with the changes predicted by the models for the impact of human activities over the next few decades. Nonetheless, the models are making steady improvements in their ability to simulate the essential features of the global climate, and continue to predict that the impact of the build-up of greenhouse gases in the atmosphere, equivalent to a doubling of pre-industrial levels of CO_2, will be a rise in the global temperature of between 1.5 and 4.5 °C.

The Hadley Centre model shows what can be achieved with current GCMs. A recent set of results[8] gives a good idea of the scale of the computations required to form a reasonable picture and the limitations of the results obtained. It also provides a measure of the latest thinking on the consequences of doubling the radiative forcing of CO_2 in the atmosphere, and of including the effect of sulphate aerosols created by the combustion of the fossil fuels containing sulphur. The latter is particularly important as it may offer an explanation as to why the warming during this century, with its temporary abatement between the 1940s and the 1970s (see Section 5.5), has not followed the course predicted on the basis of the build-up of greenhouse gases alone.

The model was brought to near equilibrium through 470 years of simulated climate. After that it was run for 300 model years under control conditions, during which there was no detectable trend in the global mean temperature. Thereafter three experiments were conducted, each starting with the conditions in the year 1860 and running up to the year 2050. They consisted of a control with constant CO_2 concentrations, an experiment in which the concentration of CO_2 is increased gradually to give changes in radiative forcing due to all greenhouse gases, and an experiment in which both greenhouse gases and the direct radiative effect of sulphate aerosols were represented. This means that by 2050 the build-up of greenhouse gases is equivalent to doubling in CO_2 levels, and the radiative impact of these dominates the climatic changes in the longer term. But the effect of sulphate aerosols is partially to counterbalance this radiative forcing, especially over India, China, Mexico and Southern Africa, and to slow global warming appreciably. The results during the twentieth century (Fig. 6.2) show a more realistic representation of climatic change during the twentieth century. The model also predicts a somewhat less substantial warming of 2.5 °C due to doubling CO_2 levels over the period 1860 to 2050, and this warming is reduced to around 1.7 °C if the build-up of sulphate aerosols follows predicted patterns.

These modelling results are a considerable step forward but they must be treated with caution. While they have combined the two most obvious products of human activities, they have not addressed many of the challenges set out in Chapter 5. In particular, the model may not be able to reflect the full extent of the natural variability of the climate or the possible alteration of the incidence of weather regimes which could play such an important part in defining future global climatic patterns. This inability to address the statistical nature of weather patterns shows that GCMs

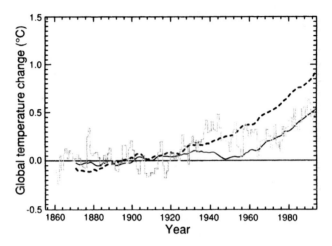

Figure 6.2. The predictions of global annual mean warming produced by the Hadley Centre, showing how the inclusion of the direct effects of sulphates (solid line) provides a better match with observations than greenhouse gases alone (dashed line). (From IPCC, 1995, Figure 6.3.)

cannot yet provide answers on how global warming will alter the loading of the climatic dice. Also, the possibility of other human activities such as deforestation, desertification, ozone depletion, and the indirect effects of sulphate aerosols exerting a significant influence has yet to be addressed.

These other factors cannot be swept under the carpet. Take the example of deforestation. We are all aware of concerns about the destruction of tropical forests, but changes in northern forests rarely intrude into our environmental nightmares. Modelling work by Gordon Bonan and colleagues at NCAR suggests this may be shortsighted.[15] They have simulated the effect of removing all the forests north of 45° N and replacing them with bare soil. Because forests with stable snow cover reflect almost half the sunlight falling on them, whereas open areas covered by snow reflect more than two-thirds of the Sun's rays, chopping down boreal forests has a cooling effect in the winter half of the year. The resultant cooling is startling. At 60° N the average fall in temperature is 12 °C, and even in late summer when there was no snow and the soil absorbed as much sunlight as the forest it replaced, the cooling was still 5 °C. So, in forming a better picture of the impact of human activities on the climate, deforestation is just one of a number of unresolved issues that will have to be addressed.

If this analysis might be regarded as too theoretical, the possible impact of soil erosion is based on more direct analysis. Since the start of the drought in the Sahel in the late 1960s there have been many observations of huge dust clouds streaming out across the tropical Atlantic each summer to North and Central America. Joseph Prospero and colleagues at the University of Miami estimate that the amount of sunlight reflected into space by this dust has had a significant regional cooling effect.[16] More generally, Inez Fung and colleagues at the NASA Goddard Institute for Space Studies, New York, estimate that the overall effect of dust formed as a result of soil erosion due to cultivation and deforestation is to produce a cooling effect comparable to that due to sulphate aerosols.[17]

These complications should not be used as an excuse for inaction. Although we have to explore the complete range of human activities, any decisions taken now must be guided by current knowledge. What we can say is the performance of current models is impressive and hence the observed warming in the last two decades is difficult to explain in terms of natural variability. So, while it is possible that by a cruel stroke of fate this work and other models have managed to produce a reasonable replica of what is known about recent climatic change, it is understandable that the IPCC has concluded[18] that 'the balance of evidence suggests a discernible human influence on global climate through emissions of carbon dioxide and other greenhouse gases.'

The next stage is to tackle the worrying discrepancies in measurements of what precisely is happening to temperatures around the world and throughout the depth of the atmosphere, and improve our understanding of how human activities are altering the climate, both in terms of the scale of warming and how this will vary from place to place. Then we must find out more about how the climate can shift of its own accord, as there is still an uneasy feeling among many climatologists that the natural variability of the climate may have been underestimated, and that GCMs also fail to simulate adequately the scale of these natural changes. These issues can only be resolved by a combined assault aimed at identifying with greater precision the match between observed changes and climate model predictions. This must be a multivariate approach, using more than surface temperature trends. It will probably rely on the three-dimensional variations in temperature, but could also exploit changes in other variables, such as precipitation, or even on the incidence of extreme weather events. Often termed 'fingerprinting', the approach relies on improving the predictions of how the impact of human activities will differ from natural variations and then showing beyond reasonable doubt what is occurring can only be attributed to these activities.[19]

In using the latest model results with improved analysis of temperature trends and spatial patterns, we must make some assumptions about the limits of natural variability. The models provide a measure of this, in that when run without any perturbations due to human activities, they exhibit fluctuations which appear to mimic natural variability well. The unresolved issue of flux adjustments does, however, cast a shadow over this approach to getting at natural variability. Where these adjustments are made, the effect may be to lock the models into a narrow view of the climate and suppress longer term fluctuations. Where the models are allowed to do their own thing they may explore the variability of unreal worlds which have little relevance to our current predicament.

An alternative approach to natural variability has been adopted at the Hadley Centre. This is to use the SST record as a measure of longer term climatic variability. A comparison is made of how the model runs differ when this SST record is combined with either a constant level of greenhouse gases, or with the increases in the levels that have occurred this century both with and without the sulphate aerosol effects. It may thus be possible to get a better handle on the anthropogenic impact on the climate. This technique can only explore the fingerprint over land surfaces, as over the oceans temperature trends are defined by SST value. This is not a problem, however, as it is over land where the models predict the biggest and most immediate effects of human activities.

The general conclusion of both the Hadley Centre work and various other efforts to detect the fingerprint of human activities on the climate is that it exists but, as yet, in a rather faint and smudged form. At ground level, the model results underestimate the recent warming over northern Eurasia and eastern North America, as the models fail to predict fully the strong westerly circulation in the northern hemisphere in the winter half of the year, which is a major factor in the warming trend since around 1970.[20] This may, however, be nothing more than the models not picking up an example of the natural variability of the climate. So part of the current uncertainty is whether much of the strong winter westerly circulation in the northern hemisphere of recent years is a temporary fluctuation, or a permanent feature of a warmer world.

In the stratosphere, the models successfully predict a marked cooling, which is part and parcel of the build-up of greenhouse gases in the atmosphere. The instrumental records since the early 1960s show a drop in temperature of around 1 °C. This is somewhat greater than predicted but the difference may be a consequence of two factors. First, many early radiosonde balloon temperature measurements are thought to be on the high side because the instrument packages were not adequately insulated against the effects of absorbing sunlight. The second complication is that the stratosphere warms sharply after volcanic eruptions (see Fig. 5.9). Since the beginning of the 1960s there have been three major eruptions (Agung in 1963, El Chichón in 1982 and Pinatubo in 1991). Although the models can do a reasonable job of predicting the impact of these important contributors to the natural variability of the climate, they are not yet able to remove entirely the blurring effect volcanoes have on interpreting the fingerprint of greenhouse gases. But, the fact that the models reproduce roughly the right amount of both surface warming and stratospheric cooling is a major reason for the IPCC consensus.

The remaining uncertainty will not be cleared up quickly. But that does not alter the urgency to improve our understanding of what the fingerprint looks like. This will involve progress on all fronts. Improved monitoring of future global climate change is relatively straightforward. Refining how models handle such factors as ocean–atmosphere coupling and the properties of clouds will prove more challenging. Exploring more fully the consequences of natural changes like volcanoes and also other human activities, including ozone depletion in the stratosphere and changes in land use, will be equally demanding. Better understanding of natural climatic variability is essential, but progress will be even harder to achieve. Nevertheless, without knowing how the climate can shift of its own accord, the problems of attribution may prove insurmountable.

These observations provide an obvious link with economic models. Alongside developing sufficient confidence in the fact that human activities are altering the global climate, we have to address questions on what action should be taken, what will be the costs of such action, and what will be the benefits. These questions require an examination of the models of the economy and whether they are capable of predicting the economic response to both climatic change and how economic systems will react to any action that is taken to prevent global warming. But before doing so, there is one other aspect of weather forecasting to address.

6.3 Seasonal forecasts and cycles

Sandwiched between these massive number-crunching exercises to improve day-to-day forecasts and to model the global climate, there exists an awkward gap of predicting seasonal weather and identifying longer term patterns. Inaccessible to direct numerical forecasting techniques which are overwhelmed by the growth of errors in the initial conditions, and too detailed to be addressed by the broad brush of the climate models, it is an area which relies on more empirical techniques than the models considered thus far. As such it combines aspects of numerical models and techniques that are more in keeping with economic models.

The best examples of this type of model have been developed to predict the development of the ENSO. Following the extreme events of 1982–83, advances on several scientific fronts have led to the capability to forecast ENSO phenomena several seasons ahead. This has transformed forecasting on such timescales which, prior to this progress, had been seen by many professional meteorologists as the preserve of charlatans. The building up of an understanding of the physics and dynamics of ENSO events during the 1980s has enabled modellers to create a variety of detailed simulations of how the atmosphere and ocean interact across the tropical Pacific. Because it is possible to exploit the relevant part of coupled atmosphere–ocean GCMs and empirical data of how past events behave, the tropical Pacific can be treated in isolation and its response to changing conditions tuned to build up a variety of different modelling schemes. These hybrid systems have proved, with varying degrees of success, that it is possible to predict the occurrence and course of ENSO events.

The performance of the various models[21] is the subject of continual review. What has emerged so far is that different models are all capable of providing useful forecasts. The best results are obtained when strong ENSO episodes occur, where these episodes included the warming of 1986–87, the cold period in 1988–89 and the rapid warming into 1990. Different models did better in different instances, but they all showed reasonable skill. But when the fluctuations were weaker or quiescent, the models were in trouble. In particular, the continued existence of a moderate ENSO warm event from the beginning of 1991 to well into 1995 caught them all out. More disturbing was the fact that the physical models did not significantly outperform empirically based models. What is not yet clear is whether the ocean–atmosphere system has sufficient inherent predictability to enable physical models to be improved appreciably beyond purely empirical models. The economic benefits of achieving such improvements is substantial because of the strength of the teleconnections between ENSO episodes and seasonal weather throughout the tropics (see Section 5.7).

In mid-latitudes the prospects for seasonal forecasts are less rosy. Although there are identifiable connections between the ENSO and seasonal patterns, they are less well established. What is more striking is that the essentially chaotic nature of atmospheric patterns in mid-latitudes appears to dominate the forecasting process. This unwelcome fact has been explored using the Hadley Centre GCM.[22] The historic data for global SSTs, going back to 1871, together with more recent sea-ice data, have been used to calculate the impact of fluctuations in SSTs on seasonal weather patterns. This work showed that these differences were inconsequential compared with the initial atmospheric conditions, which vary substantially from day to day. The ENSO had some impact in winter in the North Pacific and in summer in Europe. But the principal conclusion is that, unlike the tropics, variations between seasonal patterns from year to year are largely the product of the chaotic variations of the atmosphere. The only crumb of comfort is the hint of a signal in decadal patterns. The simulation of winters in the 1960s suggested that the prevailing SST anomalies produced an increase in the tendency for stationary high-pressure systems to form in the vicinity of the Greenwich meridian. This tallies with the North Atlantic Oscillation being particularly low during this decade (see Fig. 5.12).

These gloomy observations about seasonal forecasts provide a stepping stone into the subject of economic forecasting. The fact that empirical models can perform as well as physically more complicated approaches will, however, generate a sense of *déjà vu* with many economists. The same observation applies to the question of identifying cycles, although in this case the economic house seems to be in better order.

Regular fluctuations on the timescale from a few years to a few decades have fascinated economists and meteorologists for a very long time. In both meteorological and economic circles these 'cycles' have been the subject of detailed analysis and, where identified, intense speculation as to their causes. In the case of meteorology, the jury is still out on the question of whether any of the fluctuations, apart from the quasi-biennial oscillation (see Section 5.7), are of sufficient regularity to justify being called cycles, while their cause is of still greater dispute. With economics the case for cycles is much stronger and their causes, while a matter of debate, appear to be founded in the feedback mechanisms operating in economic systems. But, as with the case of climatic change, the ability of models to handle, and, better still, predict the occurrence, amplitude and period of such fluctuations is a major challenge for economists. So, the nature of economic cycles and whether they are predictable needs to be reviewed here.

The principal economic cycles are known by their original discoverers.[23] They are called Kitchin (3 to 4 years), Juglar (7 to 11 years), Kuznets (15 to 25 years), and, most enigmatic of all, Kondratiev long waves (around 50 years). The evidence of the Kitchin cycle was published in the 1920s and was explained in terms of overshoots and undershoots in the business inventories. As such it was an internal self-generating (endogenous) consequence of the economic system. While there is plenty of evidence to suggest that the inventory aspect of this cycle is now handled more effectively than prior to the 1920s, this periodicity is still a feature of many national economies around the world. The most obvious explanation is political: in many countries the period between elections is around four years and governments are apt to boost the economy as part of their efforts to get re-elected.

The Juglar cycle has a longer history, having been first proposed in the 1860s. It appeared to be driven by investment feedback processes of overinvesting and underinvesting in plant and equipment. Again the persistence of this fluctuation since its first discovery, in spite of increasingly sophisticated investment processes which should have damped out oscillatory tendencies, is the subject of economic puzzlement.

The Kuznets cycle is more intriguing. It is principally a feature of the US economy and shows up most strikingly in figures for immigration, construction and agriculture (Fig. 6.3). Here the explanation eluded economists until it was linked with the fact that drought in the Midwestern USA tended to occur every 20 years (see Section 2.6). This proposal was supported by analysis of rainfall in the eastern US which suggests that the 20-year cycle is a feature of this region too.[24] But tree-ring analysis in the eastern half of the country does not provide strong evidence of cycles around this period.[25] There is only a weak quasiperiodic feature at about 18 years which may be linked to lunar influences and no sign of the 22-year feature which has been found west of the Mississippi. This suggests that the climatic explanation of the Kuznets cycle in the US economy must be taken with a pinch of salt. The troughs in the late 1870s and late 1890s may have been influenced by drought in the western half of the USA. Those of the late 1910s and 1930s, although coinciding with periods of droughts, have more to do with the First World War and the Great Economic Depression. It is hard to see how these global catastrophes could have been the product of lack of rainfall west of the Mississippi.

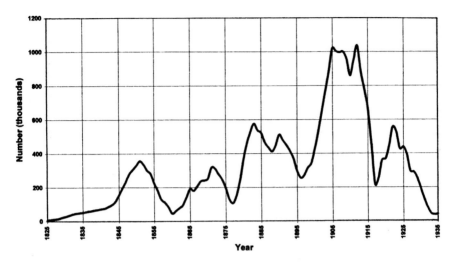

Figure 6.3. Immigration into the United States showing evidence of an approximately 20-year cycle in the number of immigrants.

The Kondratiev long wave takes us into economic battleground where prisoners are rarely taken. Usually attributed to the Russian Nikolai Kondratiev, who published his analysis in the 1920s, the existence of a long wave of about half a century in length in prices had been discussed by the British economist W. S. Jevons in the 1880s, and the Dutch Marxist van Gelderen just before the First World War. The central case was built round the rise in the general level of prices during the Napoleonic Wars, the fall to a minimum around 1850, another peak around 1870, the drop to a low level in the 1890s and another peak during the First World War. A major point of dispute in considering the significance of this periodic behaviour is the role of the global wars which marked the first and third peaks in prices. The subsequent decline in prices in the 1930s and the rises which followed into the 1970s and 1980s have only served to stoke up this debate.[26]

There is not space here to consider the reality of the long wave as a truly periodic economic feature of what are the possible causes of this behaviour. What matters is that both these long-term fluctuations and the shorter, better established economic cycles are part of the economic scene and must be included in modelling work if we are to have confidence in predictions of the impact of climatic change. So, just as the GCMs must be able to handle the natural variability of the climate, whether or not it is cyclic, the economic models must rise to the challenge of representing the regular fluctuations which have played such an important part in the economic history of the western world.

6.4 Economic models

Moving to macroeconomic models, we enter a fundamentally different world. Although both the apparatus of analysis, involving large computers and huge quantities of data, and the techniques of representing the functioning of the economy by a series of quantitative relationships appear on the surface similar, the differences deep down must be recognised from the outset. Only by facing up to these differences can we make useful links between the products of GCMs and their macroeconomic equivalents.

The first difference is that economics is not built on a set of experimentally well-established laws. Rather it is built on a set of empirical relationships and observed propensities of economic systems to behave in a given manner. While some of these relationships prove to hold good in a wide range of circumstances and over lengthy periods, others are more fleeting and ephemeral. Furthermore, conservative principles which are such a central feature of physics are a more shadowy part of economics. So, while there may be some conservative effects at work which underpin the stability of economic systems, many of the terms reflecting the inherent properties of these systems, such as 'equilibrium', 'maximisation', 'rationality' or even the 'quantity theory of money' and how to measure and control the money supply, are the subject of vigorous debate if not vituperification. This lack of a rigorous framework of laws or even agreed definitions about what precisely is to be measured makes modelling economic systems a correspondingly more inexact activity. The fact that the models rely on the maintenance of empirical relations and the extrapolating of trends to predict how the economy will perform in the future means they do not have the capability of working *ab initio* given knowledge of the current conditions and a set of laws.

There are two additional features of the empirical nature of macroeconomic models that put them in a different league. First, there is no obvious equivalence of day-to-day weather in the economies of nations. While parts of the economy, such as the stock market and commodity markets, may experience short-term squalls and eddies similar to the weather, almost all aspects of planning are concerned with how the economic climate will change over the months and year ahead. Furthermore, even on this longer timescale, many of the statistics (e.g. current account balance of payments) are subject to substantial revision months after initial publication. Such belated changes of input make the economic modellers' task far more difficult. But too much should not be read into these distinctions. Because macroeconomic forecasts are only of interest in terms of months and years ahead, they have to be considered against efforts to use GCMs to predict climatic change. The uncertainties of parameterising clouds, adjusting drift in coupled ocean–atmosphere models using flux adjustments, and the potential for sudden non-linear switches in ocean circulation put the two sets of models on a more equal footing.

The second fundamental difference is more fascinating. Economic models are part of the economic process. So forecasts prepared by governments or leading economic institutions influence how the economic world behaves. This is inevitable given how these forecasts are an integral part of policy formation and business planning. Indeed, they are used almost exclusively by governments and business for this purpose: there is economically no equivalent to the climatic studies of testing hypotheses against reality to see which theories work best. So the interaction between models and the real world is an additional complication. Depending on whether or not you subscribe to the bitterly contested theory of rational expectations, this response may be constructive or destructive. Either way it is real. No meteorologists, however bound up in their modelling work, would ever suggest that the weather is, in any way whatsoever, influenced by their forecasts.

The consequences of the interplay between economic forecasts and the economy work both ways. Most obviously, depending on how much they believe the forecasts, people will make economic decisions to alleviate or exploit the consequences of predicted changes. Conversely, there is a political pressure to include certain assumptions even though they may not be supported by experience. For instance, in the 1970s, optimistic UK government forecasts of longer term economic growth which underpinned overinvestment in the electricity supply industry (see Section 3.3) contained the assumption of the economy returning to full productive potential. This assumption was a coded way of saying that the government was committed to maintaining full employment even though it was evident that this was becoming increasingly difficult to achieve without damaging economic consequences. Events of the 1980s showed that this assumption was probably unsustainable unless it addressed more fundamental issues. The subsequent debate as to why the USA has been more successful at creating new jobs than the European Union is all part of this political conundrum. At the time it would have been political dynamite, given Mrs Thatcher's government came to power on the back of the slogan 'Labour isn't working' – a reference to the unprecedented level of unemployment in post-war Britain.

The third difference is the nature of what economists seek to predict. Almost invariably the parameters that matter to policy-makers are small residual differences between very large numbers (e.g. balance of payments or government borrowing), which are disproportionately sensitive to tiny changes in the big numbers. The difference between political happiness and misery depends on small changes in large quantities, which makes forecasting so much more difficult. Moreover, where certain outcomes are seen as political totems (e.g. low unemployment, a positive balance of payments surplus, or low interest rates), a one-sided interpretation of economic consequences may prevail and can distort or bias modelling exercises. The problem becomes even more acute if the relationship between desirable or undesirable outcomes (e.g. inflation and unemployment) appears to undergo a sudden shift after many years of relative stability.

To explore what all this means for economic models – how they compare with GCMs, and how the combination of their outputs can be used – we have to use examples. Given I live in the UK and have more direct experience of its consequences, I will concentrate on the UK Treasury model of the UK economy. This parochial approach does not limit the analysis appreciably for two reasons. First, the Treasury model is typical of a large number of economic models developed in the UK and other industrial countries by governments, academic institutions and business organisations. Secondly, we have to start at the national level as many of the empirical relationships defining economic behaviour are peculiar to each country. While many of them will be coupled to a greater or lesser extent across national boundaries by the functioning of international markets, they combine in a unique national form. Furthermore, national economies will lag or lead global business cycles and this lack of synchronicity makes it difficult to consider the global economy as more than an aggregate of national models. So, although there are times when the global economy has simultaneously to adjust some sudden change (e.g. an oil price shock), it is best to start at the national level.

Figure 6.4. Assessment of the performance of UK Treasury model comparing the prediction of GDP in the middle of the preceding year (dashed line) and the actual outturn for the year in question (continuous line). (Data provided by UK Treasury.)

The UK Treasury model has evolved over more than 25 years and has been used by the UK government and others, since it was made available in the early 1980s, to explore the possible future development of the UK economy. It grew to consist of some thousand statistical and accounting relationships which had been identified from historical data. The art of the model was to combine the various interrelationships in a consistent manner. This involves judgement both to construct a framework of the key interrelationships and to evaluate the output. The judgmental aspect of deciding what matters and of adjusting the model to produce results, that match the operators' views of the real world, past model errors and knowledge of its deficiencies, is a major factor in the performance of forecasts.

The performance of this model in the late 1970s and early 1980s (Fig. 6.4) appeared to offer some prospect that continual refinement of the assumptions included would lead to a gradual improvement of forecasting performance.[27] Instead, the process of refinement showed up the limitations of such models. The incorporation of new circumstances led to an increasingly cumbersome structure, so the limited range of experience in any economy meant that it was difficult, statistically, to distinguish between the merits of models which were internally coherent in terms of different theories. Moreover, because of the dependence on historical experience, they managed small changes reasonably well, but are all at sea if there are major shifts in economic behaviour or prices of basic commodities (e.g. large rises or falls in energy prices).

By the mid-1980s it was clear that the process of accretion in refining relationships was getting out of hand and the model received a major overhaul. The number of variables was more than halved from 1275 to 530, and 300 equations were used to describe the links between them.[28] This model came into use in June 1988, just as the seven-year economic boom was grinding to a halt. Since then the performance of the new Treasury model of the UK economy has had a tough time (see Fig. 6.4). At the same time other economic forecasts in the UK, and in other OECD countries, have been disastrous. They have failed to anticipate both the sustained downturn in the many national economies and their erratic recovery. For instance, they did not foresee the depth of the Japanese recession in 1992–93 or the strong recovery in US economy in 1992 and then its slow-down in 1993.[29] Indeed, examination of apparently widely different models shows that the differences between their forecasts are much smaller than the scale of their failures.[30] So the prospect of using an economic equivalent of ensemble forecasting (see Section 6.2) is not rosy, as clustering around 'received wisdom' is more likely to reinforce confidence in erroneous predictions than to improve assessment of future developments.

These tribulations were, in part, the product of an increase in financial deregulation in many countries. This made the timing of the changes made in the Treasury model particularly unfortunate. It highlighted the problems of trying to anticipate people's behaviour when their perceptions of economic developments undergo a radical shift. In the UK a dramatic change in public perceptions occurred in the housing market around the end of the 1980s, demonstrating the fragility of many economic relationships. In the 1980s house prices in the UK rose dramatically and home ownership rose significantly. This development was a central plank in the Conservative Government's policy and was, in part, fuelled by financial deregulation which enabled banks and other institutions to lend large sums of money. House prices boomed and many owners were sitting on huge paper profits. This impression of wealth not only promoted a sense of well-being ('the feel-good factor') but also stimulated increased borrowing to finance other consumption. The Treasury and other forecasters incorporated this economic behaviour into their models by reflecting the level of wealth in the equation for consumer spending. Given that housing represented over 60 per cent of this wealth it was important to reflect this behaviour accurately since it was influencing borrowing and spending levels. But there was no appreciable historical measurements to quantify the level of this expenditure. So different models assumed widely varying views on the size of the connection between perceived wealth and consumer spending. As fate would have it, just as allowance was being made for this increased sense of wealth, the housing market was about to go into a tailspin, in part because of decisions made by the Treasury about taxation of home loans. Within a few years the attitude to housing as a source of wealth had completely altered and all the talk was about 'negative equity'. In only five years the switch had been almost complete, from people taking too optimistic a view about their wealth to taking a far too pessimistic line.

In light of general experience, the model was slimmed to just 357 variables in 1995.[28] This move was designed principally to improve the efficiency of the forecasting and modelling work. The aim was to make the model easier to operate, develop and maintain. Other models of UK economy have undergone the similar slimming process in recent years. Particularly interesting in this process is a growing scepticism about the ability to model exchange rates and financial flows satisfactorily. So, in the case of economic models the trend is away from globalisation and towards more limited practical tools designed to meet the needs of policy-makers, business and commerce at the national level. While this approach cannot apply to a better understanding of the global climate, it had fundamental implications for the ability to model the international economic impact of climatic change.

The other important message to emerge from the track record of the UK Treasury model is how people's behaviour is shaped by their perceptions. The implications of this response for predicting the economic consequences of climatic change are profound. It is not simply that we must identify how certain aspects of economic activity (e.g. agricultural and commodity prices) may be related to, say, a warmer drier climate, but also how people may interpret the predictions and then respond to changes as they occur. This leads into the question of whether the response to policy decisions will be rational or amount to an over-reaction, possibly with perverse consequences.

At the first stage of estimating the economic consequences of climatic change it can be argued that such arcane deliberations about personal behaviour are immaterial. The models that estimate the costs of global warming, or the benefits of stopping the warming occurring by taking preventive action, do not get down to such detail. Their purpose is to provide a broader picture and in these circumstances it is wise to adopt a broadly neutral stance on the nature of people's response to the costs and benefits of global warming. While reasonable in terms of constructing manageable models, it may throw the baby out with the bath water. The historical examples cited in Chapters 3 and 4 show that even in recent times people may react in perverse ways to perceived climatic threats. When combined with the inevitable political reaction against paying too much now to avert some vaguely perceived threat far in the future, the models face an uphill task.

6.5 Modelling the economic impact of climate change

Whatever the limitations of GCMs and economic models we have to examine how they are being used to address the issues. Two areas are of particular interest. First, there is the analysis of the consequences of climatic change on specific areas of economic activity, such as world food supplies. A good example of this type of prediction is the work of Cynthia Rosenzweig, at the Goddard Institute of Space Studies, New York, and Martin Parry, while at the Environmental Change Unit, Oxford.[31] The second area is various models which have sought to address the broader question of estimating the benefit to cost ratio of taking action to prevent global warming.

The Rosenzweig–Parry model combined predictions from three GCMs for the impact of doubling CO_2 levels in the atmosphere by the year 2060. The scenarios predicted average changes of global temperatures between 4.0 and 5.2 °C – at the top end of the 1995 IPCC projections (see Section 6.2). These changes in both temperature and CO_2 levels were used to predict changes in crop yields and then used in a world food trade model to simulate the economic consequences of the changes in terms of food prices, altered trade patterns and the number of people at risk of hunger in developing countries. The major finding is that there appears to be a large disparity in the agricultural vulnerability to climatic change between developed and developing countries. Although simulated reductions in global agricultural production of major grain crops are only a slight to moderate, at low latitudes yields decline while at middle and high latitudes they rise. If these predictions are correct they have significant implication for world food policies.

There are a number of interesting features about this work. First, it draws on an immense amount of local agricultural expertise. Scientists in 18 countries estimated potential national grain crop yields using models of local production and the GCM scenarios at 112 sites. The simulations were carried out in regions representing 70–75 per cent of the current world production of wheat, maize, soybean. Rice production was less well represented, with only 48 per cent of current world production being covered. Calculations were made with and without the direct physiological effects of increased CO_2 on crop yields. This local knowledge combined a thorough scientific analysis of how crop yields would respond to technical advances over the period. The modellers were able to make a best estimate of global grain production in 2060 (some 3300 million tonnes (mt), compared with around 1800 mt in 1990) and then to predict how this figure will be altered by climatic change, CO_2 levels and changes in agricultural practice.

The second feature was the predicted latitudinal response which highlighted the potential pluses and minuses of climatic change. Near the high-latitude boundaries of production increased temperatures lengthened growing seasons and the warmth in summer increased yields. Together with the benefits of CO_2, the gains were considerable. In middle and high latitudes, where yields are already high, the additional warmth shortened the crop development period (see Section 2.2) and reduced yields. But the increased in heat and water stress levels were not significant and overall any adverse climatic effects were outweighed by the benefits of increased CO_2. At lower latitudes, however, the combination of rapid crop development plus heat and stress levels led to significant reductions in output. The model then considered how agriculture might adapt to these changes. Two levels of adaptation were identified. The first was mainly individual responses by farmers in terms of crop choices, planting dates, and irrigation levels. The second level required much greater investment in infrastructure and agricultural practice which implied intervention by central governments.

The third, and most striking, feature of the model was that the predicted changes in global grain production were relatively modest. While climatic change alone produced a reduction of production of between 11 and 20 per cent, depending on which set of GCM results was used, the inclusion of the direct effects of CO_2 and adaptation resulted in falls in predicted output down to 0 to 5 per cent with minor adaptations, and a range from plus one per cent to a fall of 2.5 per cent when major adaptations occur. The immediate reaction of those of an optimistic frame of mind is – what are we worrying about? Even allowing for the fact that the models include major assumptions about the growth in world population, how global agricultural trade will be liberalised and how prices will respond to any widening in the gap between supply and demand, these figures cannot strike fear into the hearts of anyone who believes in the power of the free market to solve economic problems. Those of a gloomier disposition rightly point to the destabilising effects of the growing gap in production and needs in the developing world while the developed world ostensibly benefits from increased production. But, this pressure can be regarded as part and parcel of the tensions that are bound to occur as the developing world seeks to achieve the economic trappings of the developed world.

Other examples of the specific impact of global warming include the consequences of rising sea levels and damage to human health. Rising sea levels have occupied a particular place in the mythology of the greenhouse effect. Maps of the British Isles showing large areas of low-lying country inundated by the end of the next century, or photomontages of great cities like London, New York or Tokyo flooded, excite public interest. Clearly, if the sea level rose uniformly by several metres the economic consequences would be massive – sufficient in themselves to justify major investment now in preventing the build-up of greenhouse gases. However, more detailed analysis of how the sea would rise in different parts of the world and revised forecasts of the average rise tell a rather different story.

A recent example of how the forecasts have been reduced come from a panel of experts advising the US Environmental Protection Agency (EPA). In 1983 it forecast a rise of 175 cm by the year 2100.[32] By 1995 this figure had been revised downwards to 34 cm.[33] A similar, if less dramatic, reduction has occurred in the IPCC Reports. In 1990 the range of projections by 2100 was from a high of 110 cm to a low of 31 cm with a median value of 66 cm. In the 1995 report these figures had roughly halved to 49, 17 and 34 cm.[34] The reduction was almost entirely the result of the lower estimates in the scale of global warming by 2100. Possibly more relevant in respect of public perceptions, the majority of this rise will be the thermal expansion of the oceans, and the melting of glaciers and minor ice caps. The melting of the Greenland and Antarctic ice sheets will have a negligible effect over the next century, as any peripheral ablation is expected to be compensated by increased precipitation. So, while the press is full of stories of huge icebergs being calved from Antarctica, which may or may not be confirmation of global warming trends, this is of no direct relevance to sea levels in the next few decades. The popular image of the rapid disappearance of the great ice sheets appears to be a much more remote prospect.

Even if the warming reaches values where the Greenland ice sheet does start to collapse, the changes in sea levels will be complicated by how the Earth's crust responds to the reduced load. As the ice melts, it will rise in the region of Greenland and around the North Atlantic. In Britain, if the partial melting of the Greenland ice sheet was sufficient to raise sea level by a metre globally, the sea level in Scotland would fall, and rise by no more than 23 cm in southern England.[35] While melting of the Antarctic ice sheet would have a more uniform impact, this appears to be an even more distant threat. So the latest thinking is that sea levels will rise slowly throughout the next century unless early action is taken to reduce the build-up of greenhouse gases. But the images of large scale coastal inundation due to the melting of the ice sheets is much farther off. Nonetheless, if the rise in sea levels is accompanied by an increase in storminess in certain parts of the world, then frequency of damaging floods is liable to increase, especially if population and economic pressures lead to the increased occupation and exploitation of the vulnerable areas.

The forecast of increased health hazards has focused on both the spread of tropical diseases and the consequences of more frequent heatwaves in urban areas.[36] Malaria has been the subject of particular analysis. At present over two billion people in the tropics and subtropics are at risk and some 270 million are infected each year. Global warming could lead to considerable expansion of the range of this disease. Where it is endemic people develop immunity and most deaths are among the very young. Major epidemics are rare. But on the fringes of its terrain where it is present for only a few months each year it can do much greater harm. Moreover, where it extends into new areas for the first time it may lead to death rates as high as 20 per cent. Various climatic models have been used to estimate how a warming due to a doubling of CO_2 levels could extend the range of malaria. They produce figures of up to a 30 per cent increase (17 million km^2) in the area where it would be endemic and an increase of over a half (some 25 million km^2) in the range of seasonal malaria. This extension could encompass many parts of Europe, including Britain, Russia, North America and Australia which are not already susceptible to the disease. Such changes could lead to 50 to 80 million new malaria cases per year.[37]

Estimates of the impact of more intense heatwaves depend on the increased incidence of temperatures above certain levels. As daytime highs rise above 35 °C (95 °F) the excess mortality in many mid-latitude cities rises sharply, especially if accompanied by high night-time minima. If average summer temperatures rise by 2 to 4 °C, the excess mortality each summer in cities like Chicago or New York could rise by several hundred. The widespread use of air-conditioning will tend to keep these figures down, but in developing countries the rises in mortality in cities like Cairo and Shanghai will be even greater. The fact that, apart from global warming, other forms of local pollution will make cities even more unhealthy places in heatwaves can only serve to compound these problems.

Estimating the costs of reducing the build-up of greenhouse gases in the atmosphere is the preserve of top-down models.[38] These seek to simulate the global economy in a manner that identifies the principal sources of emissions of these gases and the economic consequences of reducing their production. In most cases this boils down to reducing CO_2 emissions, and often to considering the implications of a carbon tax which would be specifically aimed at penalising these emissions. The significance of this tax is that it would recognise that for unit output of energy the combustion of coal produces some 80 per cent more CO_2 than natural gas while oil produces 40 per cent more than natural gas. The fact that other gases including chlorofluorocarbons (CFCs), methane and sulphur dioxide also contribute to the greenhouse effect is incorporated in some models. These models use broad empirical relationships about the sensitivity of the economy to energy availability and do not address the question of how specific technologies might tackle the problem of CO_2 emissions. It is generally accepted that such a 'bottom-up' approach, which seeks to aggregate the impact of a host of different initiatives and technologies, contains too many unknowns to provide useful guidance.

The models consider the reduction of carbon emissions by five means:

1. intra-fossil fuel substitution (IFFS) which replaces coal by natural gas;

2. non-fossil fuel substitution (NFFS) which replaces fossil fuels by non-fossil fuels (e.g. nuclear, renewables etc.);
3. substitution of energy by labour and capital in the production process (other-factor energy substitution (OFCS);
4. substitution of energy-intensive products by non-energy-intensive products in the consumption mix (product substitution); and
5. reducing deforestation or increasing afforestation.

Broadly speaking, various models address the issue that unrestrained emissions of CO_2 to the atmosphere could rise threefold by 2100. This means that during the twenty-first century the average emissions would be twice current levels. But to stabilise atmospheric concentrations of CO_2 at roughly current levels will require these emissions to be cut by two-thirds. This does not mean that energy consumption has to be cut as drastically, as the scope for moving to non-fossil fuels and replacing coal by natural gas might be able to absorb half this shift. Furthermore, substitution by other factors (OFCS) and production substitution could absorb additional reductions, so at worst energy consumption might need only to fall by about a third and this figure could be considerably lower. Changes of this order may not be asking too much as between 1972 and 1987 the effect of oil price rises in the USA meant that carbon emissions did not rise while the GDP rose 46 per cent in real terms. There has, however, been a 10 per cent increase in the consumption of fossil fuels in the USA between 1987 and 1994.

The validity of any of these top-down models does, however, depend on the realism of the scope for substitution in forms of energy production. If these alternative forms of energy supply cannot fill the gap created by cutting carbon emissions, then the cutbacks in energy consumption will have to be correspondingly greater. But even if the reduction in energy consumption had to be a third in advanced industrialised countries, the impact might be relatively modest. This is because the energy sector amounts to only about 6 per cent of GDP. So, providing this reduction did not damage the productive capacity of the rest of the economy, but was absorbed by the energy sector, then it would only amount to a 2 per cent cut in GDP. For developing countries the situation may be rather different.

As to how these reductions might be achieved, the models most frequently use a carbon tax to reduce CO_2 emissions to the required level. Figures typically in the range US$100 to 400 per ton are used. If these figures look terrifying, it is worth remembering that $250 per ton is equivalent to $0.75 per gallon of gasoline or $30 per barrel of oil. While the latter figure is high, in real terms, it is now the same order as the prices that were absorbed around 1980 during the early stages of the Iran/Iraq War. As for a tax of $0.75 per gallon on gasoline, this may send shivers down American spines, but in Europe this figure is modest compared with current taxes. So it is reasonable to assume that if there is sufficient political will these levels of taxation could be adopted as part of a strategy to prevent the unacceptable consequences of global warming.

The development of this political will depends, among other things, on minimising the economic damage of such a tax. In principle, if a carbon tax was to be used to raise government revenues instead of other forms of taxation, then the economic losses would be much less than the absolute figures might suggest. Everything would depend on the compensating economic effects of the reduction of other taxes and on the ability to substitute for more expensive fossil-fuel energy and whether this combination would reduce the demand for goods and services. In modelling these processes a central assumption is whether both the reduction in carbon emissions and the economic costs of imposing a carbon tax will be linearly related to the level of tax or whether it will be a non-linear response. In the latter case the economic costs would rise disproportionately faster than the taxation levels, because the amount of taxation required to produce a unit reduction in carbon emissions would have to rise to achieve each successive increment of cutback.

Given the emphasis on the non-linear behaviour of both the climate and economic systems so far in this book, it is hardly surprising that I will concentrate on this response. This may lead to cries of 'foul' among those who support the conclusions of linear models, because it loads the dice against taking aggressive action on abatement and favours an incremental approach. This is understandable, but virtually all the examples presented here show that economic systems respond more predictably to small changes than to sudden shocks. So unless we are confronted with incontrovertible evidence that the economic costs of climatic change are of such disastrous proportions that immediate dramatic action is required, it is probably safer to assume it is better to take small steps rather than great leaps forward. But, as will be seen, this implicit assumption on non-linear behaviour does not, however, lead to fundamental differences in the estimates in the costs of global warming in the next few decades.

The various economic models have been surveyed in detail by William Cline.[38] The striking feature of this survey is that, in spite of considerable differences in both quantifying the economic relationships and aggregating various regions around the world, the global costs of reducing carbon emissions are broadly of the same order. The reduction in GDP associated with halving predicted carbon emissions by around the middle of the next century falls in the range one to five per cent. Cuts of three-quarters by 2100 may only involve reductions of between two and four per cent in GDP. The majority of these models predict that these cutbacks can be achieved by applying carbon taxes in the range $100 to 400 per ton.

The difference between linear and non-linear models shows up most in the early stages. The leading protagonist of the non-linear approach, William Nordhaus,[39] predicts that initial reductions can be achieved at little or no cost, but that beyond around a 25 per cent cut in carbon emissions the costs steepen rapidly, and the benefits of such action become increasingly questionable. Even so he estimates than an 80 per cent cut in predicted emissions by 2100 would only involve around a four per cent reduction in GDP.

Hidden within these global figures is a major political snag. This is the uneven impact on different parts of the world. The models suggest that the developing would, and China in particular, will have to shoulder a much greater part of the economic burden. When combined with the fact that China and India are predicted to become major contributors to the sulphate aerosol burden, which may be partially offsetting the warming effect of the build-up of greenhouse gases, it is evident that difficult negotiations will be involved in sustaining international agreements to reduce carbon emissions. Much will depend on the perceptions of different countries of what will be the impact of changes in regional climate patterns (e.g. modification of the summer monsoon in southern Asia – see Section 7.4).

Whatever the course of international negotiations, two interesting features emerge from these model results. The first is the relatively modest level of the costs that might be incurred in reducing carbon emissions. The second is the fact that the abatement costs are of the same order as the estimates of damage caused by global warming. Recognising the huge uncertainties in both sets of forecasts this a particular dilemma for governments in deciding when and how much action to take. At best the arguments are finely balanced – a recipe for procrastination. But before trying to reach conclusions about whether this is the right approach, there is the interesting issue of how forecasts will drive events, whether or not politicians and the general public like it or not.

6.6 Notes

1 See Chapter 1 by S. Schneider in Trenberth (1992).

2 Palmer (1989).

3 Lorenz (1982).

4 Palmer (1993).

5 Trenberth (1992), p. 18.

6 IPCC (1995), p. 261.

7 Ibid., p. 237.

8 Mitchell, Johns, Gregory & Tett (1995).

9 A large number of papers have been published on this subject and it continues to be fiercely debated. The state of play is succinctly reviewed in Wiscombe (1995).

10 Cess *et al.* (1995).

11 Kerr (1995).

12 Tony Slingo (private communication).

13 Cess *et al.* (1991).

14 IPCC (1995), pp. 240–4.

15 Bonan, Pollard & Thompson (1992).

16 Li *et al.* (1996).

17 Tengen, Lacis & Fung (1996).

18 IPCC (1995), p. 5.

19 Ibid., Chapter 8.

20 Chris Folland (private communication).

21 Barnston (1995).

22 Folland & Rowell (1995).

23 Tylecote (1991).

24 Currie (1988).

25 See Chapter 17 by Stahle & Cleveland in Bradley & Jones (1995).

26 The differing views on the reality of Kondratieff waves are explored in great detail in Solomou (1990) and Tylecote (1991).

27 Burns (1986).

28 Chan, Savage & Whittaker (1995).

29 Ormerod (1994), p. 104.

30 Ibid., p. 105.

31 Rosenzweig & Parry (1994). (This work has been refined and expanded to form Chapter 2 of Strzepek & Smith (1995), but these changes do not materially alter the conclusions reached here.)

32 Hoffman, Keyes & Titus (1983).

33 News item, *New Scientist*, 4 January 1995.

34 IPCC (1995), Figs. 7.9 & 7.10.

35 Parry & Duncan (1995).

36 See Chapter 5 by Kalkstein & Tan in Strzepek & Smith (1995).

37 Stone (1995).

38 Cline (1992).

39 Nordhaus (1990), (1992).

7

Consequences of forecasting

'When *I* use a word,' Humpty Dumpty said in
a rather scornful tone, 'it means just what I
choose it to mean – neither more nor less.'
Through The Looking-Glass, Chapter 6

Behind all the analysis of the economic impact of climatic change there lies the basic issue of whether forecasting is a good thing, or whether it does more harm than good. One thing is certain, forecasts are bound to be produced. So what matters is how much effort society devotes to this work and how its results are integrated into the most productive management of our affairs. So far we have taken the fatalistic view that there is no alternative to using forecasts as the option of disregarding them altogether is politically unacceptable. But we should be able to do better than this. It would be reassuring if, on balance, it could be shown that there are measurable benefits in forecasting. As forecasts are such an integral part of so much planning, it would be worrying if this was not the case, at least in the case of predictions which enable us to avoid short-term extremes and potential disaster.

In parallel with the simple question of whether forecasts are worth the effort, there is the question of the less quantifiable ways in which they infect all aspects of our lives. Although many people have a healthy scepticism about all forms of prediction we cannot get away from the profound influence they have in either making decisions, or, at least, justifying them. This means that, in the longer term, measuring the benefits may be less easy to determine. How much of the costs of overinvesting in the UK electricity supply industry in the 1960s and 1970s can be laid at the door of optimistic forecasts as opposed to political pressures to show commitment to full employment and develop the nuclear power industry? Once we get into the interactive world of predicting sustained fluctuations, which influence the prices of commodities and alter property values, measuring the economic consequences of forecasts becomes even more difficult as they have major implications for the operation of the markets. Between the extremes of the forecasts being regarded as wholly unreliable and being believed implicitly, there lies a fascinating continuum of responses from sober investment to cynical speculation, much of which cannot be quantified. This range is best explored by analysing how various forecasts are used and what the consequences appear to be.

7.1 Weather warnings

With short-term forecasting the analysis is relatively straightforward. Many studies have shown that in the developed world improved forecasts of severe weather have delivered major benefits in terms of reduced loss of life and improved operation of emergency services. The improvements are evident in many examples ranging from tornadoes and hurricanes, through snowstorms to major flooding episodes. Broadly speaking, there has been a tenfold reduction in deaths associated with such events during this century. The parallel rise in costs of damage to property is a separate matter. This is not related to short-term forecasting, even though it is the product of intense, and often predictable, events. Because only a tiny part of the damage can be prevented at the time by, say, boarding up windows, the real issue relates to much longer-term investment decisions. These involve the design and siting of property to make it less vulnerable to extreme events, and whether these events are becoming more frequent, which depends on accurate forecasts of climatic change.

Where the claimed benefits of short-term forecasts need to be the subject of closer scrutiny is in identifying the wider economic consequences of these forecasts. Clearly, a wide variety of industries, notably agriculture, construction, transport and the offshore industry, rely on standard weather forecasts to plan their operations. Weather services have estimated that the savings exceed the costs of preparing the forecasts by at least a factor of ten. What is less clear is whether, once these services are factored into their operations, both individuals and businesses do not sail closer to the wind. This process is further complicated by shifting perceptions of risk and experience of the accuracy of the forecasts. Growing confidence in forecasts may lead to riskier behaviour, whereas one false alarm when, say, a hurricane shifts its path in the last few hours can have a major impact on how future warnings are heeded. So, while in the normal course of events the accurate forecasts do provide effective warnings which prevent disruption and have real economic benefits, the costs of being caught out by extreme or unexpected events may be consequently all the greater.

7.2 Weekly weather

When it comes to medium-term forecasts ranging from a few days to a couple of weeks, the focus of the analysis shifts. Advanced warning of extreme events can be helpful in getting emergency services ready, although these plans are bound to be refined in terms of subsequent forecasts. Where they assume greater importance is in planning operations which involve stocking up to handle sustained weather demand, or planning weather-dependent activities. So retailers of perishable foods such as salad vegetables exploit accurate forecasts of hot weather, farmers plan activities, such as sowing and reaping to take advantage of the best conditions in the days ahead, and offshore operators schedule hazardous operations to avoid bad weather. But, as with short-term forecasts, errors can generate adverse reactions. A too gloomy forecast of the weather over, say, a long holiday weekend will produce howls of anguish from the tourist industry if people are frightened off and then the weather is better than expected.

An appropriate measure of the economic benefits is the success of marketing specialised medium-term services to meet the needs of various industries. While national weather services provide extensive information about their forecasts up to a week ahead, interpretations of how the predicted conditions will affect specialised operations requires additional analysis. The fact that many major businesses finance these services is the clearest measure of their economic value to their users – on balance they are earning their keep. It also confirms both the quality and the benefits of microeconomic analysis and forecasts. At the level of a few days to a couple of weeks, predictions of abnormal weather can be accurately translated to changes in demand for specific goods or services. It is when it comes the longer term consequences of forecasting, and their macroeconomic consequences, that the implications become more complex.

7.3 Seasonal forecasts

The simplest example of the complicated consequences of longer term forecasts was identified in Chapter 2. This is the anticipation of good or poor harvests. The extreme example of the positive aspects of this is the biblical example of Joseph foreseeing the seven years of plentiousness in the land of Egypt and the seven years of dearth. By building up grain stocks in the good years Egypt was able to have bread in the bad years when the surrounding lands experienced famine. A more typical response was seen in the famine of 1315–16 when rich merchants built up hoards of grain as it became evident that the harvest would be inadequate, and, in due course, the poor starved. This ability of a minority to exploit forecasts at the expense of the majority is an inevitable feature of free markets. The solution to exploitation of information is to make it widely available so that the market can operate more efficiently. Unfortunately, if predictions that a commodity will be in short supply at some time in the future are widely believed, they can lead to panic buying which drives prices to exceptional levels. The example of the frosts in Brazil in 1975 which destroyed much of the coffee crop had precisely this effect. Not only did market prices rise sharply as the scale of the damage sank in but also shoppers stripped the stores bare to build up personal stocks, thereby increasing demand and further inflating the prices. The peak of over $3.70 per lb in early 1977 has not been exceeded since, in spite of droughts in the 1980s and damaging frosts in 1994.

In spite of the fact that these distortions are bound to be a consequence of forecasts, the impact of seasonal and annual patterns in the weather on the harvest is so great that the benefits of improved predictions are potentially huge. In particular, the possibility of developing seasonal forecasts in the tropics linked to predictions of the ENSO (see Section 6.3) contain the prospect of offering real benefits in both choice of crops and planning a more measured response to poor harvests to minimise the impact of shortages. These benefits are more readily identifiable because they can be related to fluctuations in a single variable: rainfall during a specified wet season. By comparison, the impact of extreme seasons in mid-latitudes is often linked to a more complicated mixture of meteorological parameters or to fluctuations in rainfall over longer periods (see Section 2.2).

The potential benefits in the tropics are already being recognised. In Peru the possibility of predicting ENSO events has already been incorporated in national planning of the agricultural sector.[1] Governmental and intergovernmental organisations are trying to influence farmers' decisions on planting either cotton or rice in northern coastal areas where rainfall amounts vary dramatically between ENSO events and years when the waters off the coast are colder than average. Because rice requires plentiful rainfall it does best in El Niño years, whereas cotton is more drought tolerant and does best when the reverse conditions apply. A measure of this approach is that during the 1986–87 ENSO the overall output of the agricultural sector remained roughly steady as compared with the dramatic fall during 1982–83. A similar success was achieved in north-east Brazil in 1992, which was also an El Niño year. Rainfall in the region was approximately 70 per cent of normal, about the same as in 1987, when no El Niño related action was taken and grain production was less than 20 per cent of normal. In 1992, El Niño was predicted and drought-resistant crops were planted, with the result that grain production achieved a much higher figure of nearly 80 per cent of normal.

The fact that the forecasting of ENSO events does best when predicting extreme conditions is not a particular drawback in respect of agricultural planning. What matters most is to pick out accurately the very wet and very dry seasons when the choice of crop really counts. Providing less extreme seasons are also correctly identified without knowing whether they will be on the wet or dry side, then choices matter much less.

Elsewhere in the tropics forecasting the ENSO has potential application in many countries. The high correlation of maize yields in Zimbabwe with the ENSO (see Section 4.5) permits authorities to anticipate good and bad harvests. In the same vein, the UK Meteorological Office has been conducting forecasting work on seasonal rainfall in both north-east Brazil and the Sahel with considerable success. The performance of these forecasts and the implications for managing both agriculture and aid to sub-Saharan regions has been examined in detail.[2] The combination of meteorological research and economic analysis provides important insights into the challenges of producing forecasts which are of real value to developing countries.

At the meteorological and the climatological level the results are impressive. Two approaches have been developed to predict rainfall during the wet season from July to September. The first uses a set of statistical relationships between sea surface temperature anomalies (SSTAs) throughout the tropics not just ENSO events, and rainfall anomalies across the Sahel. The latter are defined in five discrete categories (very dry, dry, average, wet and very wet), each of which is equally likely to occur over a long period. These rules have been used to produce forecasts which show considerable skill using April to May SSTAs. But they cannot yet predict precisely where the rain will fall, and when. The research does show, however, that there is a prospect of being able to produce skilful forecasts of detailed rain patterns.

The other approach has been to explore how the output of a GCM handles the rainfall across the Sahel as a function of SSTAs. This work has explored the performance of the model using historic data. It showed that the model had the potential to produce good forecasts of seasonal rainfall providing it had June SSTAs, but this dropped off sharply with data from April. This fall in skill as the lead time increased to a few months was much larger than in the statistical method. This is a major shortcoming as forecasts in July are far too late to influence planting decisions.

Examining the implications of using these forecasts to plan activities in the Sahel produces a more demanding set of criteria. These depend on who is using the information. The users range from international aid organisations, through various government organisations in the different countries to the farmers, pastoralists and freshwater fisherman who are most affected by the rainfall fluctuations. Starting with those who have most to gain from better information, clearly farmers need forecasts in good time to make decisions about which crops to sow, which requires a reliable forecast by early May at the latest. This information then has to be made readily accessible to producers to influence sowing decisions. Thus rainfall forecasts need to be reduced to a form which can be broadcast by radio, reflects local calendars, fits in with other factors influencing agriculture, and stresses the strengths and weaknesses of what is being predicted. To meet these criteria adequately it must also be relevant to local conditions and recognise local knowledge otherwise it will not affect producers' actions. Hardly surprisingly, current forecasts fall well short of these demands, and in the foreseeable future it is likely that only large commercial growers with irrigated land are likely to benefit from the predictions that will be developed. At the other end of the scale the poorest producers, who lack the flexibility to respond to forecasts, are likely to become more vulnerable to how the markets respond to the information and in greater danger of impoverishment.

For donor agencies and governments the potential to exploit the forecasts is greater. Again the information must be timely. At the most mundane level, donor agencies have to allocate their aid resources early in each financial year. So here again reliable forecasts are needed early in the year, especially for organisations whose financial year is the calendar year (e.g. EU Commission, France and Germany). The non-linear response of agriculture means in providing aid and developing national responses to extreme years what really matters is getting accurate early predictions of very dry or very wet years. By comparison less extreme years are of little concern, providing their ordinariness is correctly forecast. The benefits of forecasting an exceptionally dry season show up not only in planting decisions but also livestock management. The ability to draw up a national plan of herd de-stocking in a staggered manner to avoid the all-too-common collapse of livestock prices and a glut on the market in times of crisis could completely alter the lives of those dependent on herds for their livelihood. These benefits would be greatly enhanced if forecasts could identify regional variations in rainfall; national authorities could then anticipate transport and distribution pressures to even out food balances and manage stock movements.

7.4 Monsoon forecasting

The implications of developing and using seasonal forecasting of rainfall can be seen also in the lengthy history of efforts to predict the summer monsoon across the Indian subcontinent. Following the famine in 1877 caused by the failure of the monsoon that year, the Government of India called upon H. F. Blandford, the Chief Reporter of the Indian Meteorological Department to prepare monsoon forecasts. Blandford concluded that 'the varying extent and thickness of Himalayan snows exercise a great and prolonged influence on the climate conditions and weather of the plains of north-west India'.[3] This relationship suggested that deep snow cover during the spring would lead to a dry summer, whilst a thinner cover was associated with wet seasons. Initial forecasts between 1882 and 1885 encouraged Blandford to start operational forecasts covering the whole of India and Burma in 1886. Since then, predicting the summer monsoon has been an important task for the Indian Meteorological Department.

After 1895, under Sir John Elliot, the forecasts took account of not only Himalayan snow cover, but also peculiarities in pre-monsoon weather across India and over the Indian Ocean and Australia. This wider perspective was developed by Sir Gilbert Walker, when he was Director General of the Indian Meteorological Department, in his ground-breaking work on global scale oscillations (see Section 5.7). This established the importance of the Southern Oscillation to the development of the summer monsoon, and put the whole forecasting process on a more formal footing. It is a measure of the importance of this work that between the 1920s and the early 1980s the forecasting methods made little progress. Although new predictors were identified and some of the established relationships fell from favour (the connection between the monsoon and Himalayan snow cover appeared to shift and was dropped from forecasts in the 1950s) the general performance stagnated. So, although the forecasts have been a valuable part of the planning in India for many years, the lack of progress was frustrating.

Studies since the early 1980s have shown that changes over time in empirical relationships between indicators around the world and the subsequent monsoon are an important factor in developing improved forecasts. One problem was the quality of the data. In the case of Himalayan snow cover recent satellite observations have revived interest in this predictor. They show that the relationship first identified by Blandford is a useful guide, but that the extent of Eurasian–Himalayan winter snow cover was a better indicator, given the geographically uneven and variable nature of snow cover over the Himalayas, Tibet and Siberia. In addition, the role of the ENSO has been established more clearly. One reason for this is that since the 1950s the ENSO seems to have played a more dominant part in both global climate fluctuations and the monsoon, with warm El Niño events preceding dry summers and cold La Niña events, when the eastern equatorial Pacific experiences well below normal temperatures, producing plentiful rainfall. The severe drought in 1972 (see Section 4.4), which cut the harvest in Maharashtra state by more than half, is the clearest example.

The examination of the various factors influencing the strength of the monsoon provides many insights into global circulation patterns. In developing a forecasting process, 19 significant predictors have been identified, many of them closely interrelated. As well as the indicators already mentioned, and various local parameters, these predictors also include northern hemisphere winter temperatures and the quasi-biennial oscillation (QBO – see Section 5.7). Selections of these predictors can deliver impressive forecast records (Fig. 7.1). But the optimum statistical combination varies from time to time. For instance, although El Niño events are normally linked with dry years in India, there have been periods when

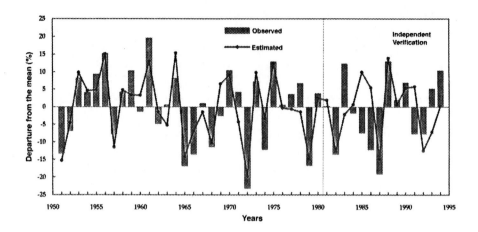

Figure 7.1. The comparison of a statistical methodology using 19 predictions to forecast of the all-India monsoon summer rainfall. (Reproduced by permission of the Royal Meteorological Society.)

this has not been the case: the prolonged event from 1991 to 1994 did not result in severe drought in India.[4] Moreover, as with forecasts of ENSO itself (see Section 6.3), its predictive value in respect of the monsoon falls apart when the El Niño is weak.

This experience yet again highlights the limitations of using statistical formulations to produce seasonal forecasts. Moreover, in spite of the lengthy experience of refining the connection between the monsoon and various precursors, the compilation of more and more information has produced only limited improvements. The gaps in the data are part of the problem, but the absence of an accepted physical explanation for the connections is a far greater drawback. In this context, parallel work on using GCMs to study what leads to wet and dry seasons has produced disappointing results. While they can reproduce the interannual variations in monsoon circulation in response to tropical SST anomalies the simulations of interannual monsoon rainfall fluctuations are very poor.[3] From the examples of the very dry year 1987 and the very wet year 1988, the bias of the GCM simulations was for more rain in the latter season, the results being extremely sensitive to the initial atmospheric conditions. This chaotic response mirrors the simulations of mid-latitude seasonal circulation patterns (see Section 6.3).

The limitations of empirical methods and the evidence that GCMs still have a long way to go before they can produce operational seasonal fore-

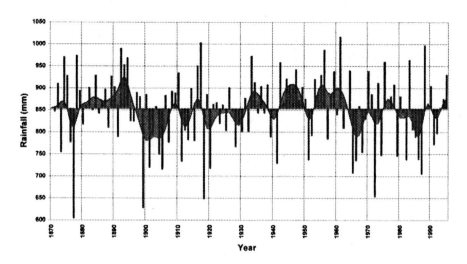

Figure 7.2. Variations in the all-India summer monsoon between 1871 and 1994, showing no significant increase in recent decades, together with smoothed data showing longer term fluctuations.

casts has important climatic implications. In terms of the direct exploitation of current forecasts the same challenges identified for the Sahel apply here. The empirical rainfall forecasts lack the detailed local predictions which, if believed by farmers, could influence behaviour. At the levels of regional and national the forecasts have greater potential, notably in terms of planning disaster relief.

As for the GCMs, the implications are more subtle. While their performance in terms of seasonal forecasting does not directly undermine their predictions of climatic change, it does underline the problems of being able to predict the changing frequency of extreme seasons. This is essential not only to understanding how climatic change will affect many economic activities, but also in convincing national governments that the specific consequences will be damaging for their own countries. So in the case of India, confident forecasts that global warming will lead to, say, a quantifiable increase in dry monsoon seasons would have a much greater impact on policy formulation on carbon emissions than simply stating that temperatures will rise a couple of degrees. Against a background of there being no significant trend in monsoon rainfall across India since 1870 (Fig. 7.2), nor any marked increase in the incidence of wet or dry seasons, only convincing forecasts are going to dispel a natural propensity to wait and see.

Current GCMs are equivocal about the impact of global warming on tropical rainfall. On balance they tend to point towards a stronger Asian summer monsoon,[5] but there is considerable debate about the spread in the results. Some of the more recent ocean-atmosphere GCMs hint at the possibility of tropical desiccation with a weakened summer monsoon in a warmer world. The inclusion of the effects of sulphate aerosols appears to be a crucial factor in this switch. The debate is enlivened by a palaeoclimatological controversy. It has long been believed that during the last ice age the tropics did not cool down anywhere near as much as at higher latitudes. This tropical stability added weight to the conclusions of GCMs which predict that polar regions will warm far more than low latitudes. It also reinforced the argument that it would be the developed nations who would have to put their house in order as they stood to lose more with global warming. More recent studies of the ice age climate suggests that the tropics may have experienced changes comparable to the rest of the globe.[6] So global warming could pose every bit as much of a threat to the developing world. These differences need to be resolved, because India, China and their neighbours are bound to play a central role in any action to reduce the growth in carbon emissions in the twenty-first century.

7.5 Hurricanes

Another area where the identification of teleconnections across the tropics is producing increasingly useful seasonal forecasts is in respect of hurricane activity in the North Atlantic and Caribbean. Developed by William Gray and colleagues at Colorado State University, these forecasts are built round the strong association between rainfall in the Sahel region of West Africa and the number of intense hurricanes that hit the United States.[7] The hurricane season in the Atlantic starts in earnest at the beginning of August and runs to around the end of October. The forecasts use three empirical forecasting routines.[8] The first is produced at the beginning of December for the next year, the second is on 1 June and the final one is on 1 August as the season gets going. Each one uses a different but related set of predictors.

Nine months ahead the forecast is built round the QBO, the rainfall in West Africa in the previous summer and the latest ENSO forecast (warm episodes tend to reduce hurricane activity, while cold events enhance it). The inclusion of the QBO is an interesting addition. This oscillation is not only reasonably regular in its 27-month period, but also the reversal of the winds first takes place high in the stratosphere and propagates downwards over several months; so its effects can be anticipated with considerable accuracy. Furthermore, the fact that the change in direction of stratospheric winds can influence the weather in the lower atmosphere by either enhancing or suppressing hurricane activity may be the key to other seasonal forecasts.

By June the forecast is refined to include current ENSO conditions rather than forecasts. It also takes account of pressure patterns and high level winds over the Caribbean and temperature and pressure patterns over North Africa north of the Equator. The final forecast is able to use information extending right up to the start of the hurricane season. The most important addition is actual rainfall in the western Sahel in June and July.

All three forecasts show considerable skill in predicting the number and intensity of hurricanes during each season. They do not, however, give any guidance as to when the storms will occur and what paths they will follow: this remains the preserve of the numerical weather forecasts. Nevertheless the relative success of the December predictions is particularly rewarding, as it suggests that useful warnings of active hurricane seasons can be provided in good time. While the forecasts in June show considerable improvement, the August ones have yet to show significant additional benefit.

The future success of these forecasts depends on whether the general behaviour of hurricanes in the tropical Atlantic alters. Gray considers that the decline in activity in recent decades is linked to a weakening of the thermohaline circulation in the Atlantic (see Section 5.8). There is some evidence of the first signs of this process speeding up again and this may lead to increased hurricane activity in the years ahead. So the forecast for 1996 included an analysis of the likely increase of major hurricanes striking the US East Coast, Florida and the Caribbean basin.[9] It concluded that a multi-decadel circulation change could be in progress with the ominous prospect of an increase in these hurricanes. At the same time rainfall in the West African Sahel would increase. The active hurricane season in 1996 suggests these proposals might be on the right track.

Possibly of more immediate relevance is recent work on decadal oscillations of SSTs in the tropical Atlantic. This shift of areas of above and below normal temperature between 15°N and 15°S has a period of about 13 years and seems to be the product of atmosphere–ocean interactions, which have been successfully simulated in a computer model.[10] So there is the prospect that this particular phenomenon can be incorporated into improved long-term forecasts of both Atlantic hurricane activity and Sahel rainfall.

7.6 Extratropical seasonal forecasts

The prospect for producing seasonal forecasts in mid-latitudes that perform as well as tropical efforts is remote. The economic arguments for developing such forecasts are, however, substantial. There is also an interesting political reason for producing success in this area. Seasonal forecasting is seen as a haven for climatic change research in the increasingly stringent funding climate in the USA. The supposition that the potential economic benefits of seasonal forecasting will provide a financial umbrella for climatic work may prove to be illusory. If such forecasting is as difficult as has been suggested here, then the forecasters' paymasters may soon grow weary of waiting for promised success. So the forecasters must cross their fingers that either the research makes more rapid progress than expected, or that they strike lucky with their early predictions. The forecast of a mild winter across the south-eastern third of the USA for December 1995 to February 1996 together with above average rainfall in the central tier of southern states shows what they are up against. The winter in the south-east was cold, the central part of the country was very dry, especially Texas, and the near record-breaking warmth in the south-western states completely eluded the forecasters. Fortunately, the performance in the winter of 1996–97 was rather better.

As for the economic benefits of seasonal forecasts there is an additional problem of what proportion of the costs of extreme seasons are avoidable. In the tropics and subtropics, where such forecasts show the greatest skill, the need for considerable improvements in performance have been identified. Here again the key is better physical understanding of what drives both interannual variations and establishes the regional patterns which define differences in local rainfall. Only when these issues have been unravelled so that farmers can take informed decisions in good time will the real economic benefits be achieved. While the benefits of forecasts using existing methods for guiding wider planning may repay the costs of this work, they represent only a small part of what might be achieved.

Consideration of the potential benefits of seasonal forecasts for developed countries in mid-latitudes, in cash terms, shows that the potential for savings is greater because of the wider range of economic impacts that extreme weather can have on many aspects of life. Although, as a proportion of GDP, these costs are much smaller than those resulting from harvest failures in developing countries their varied nature seems to offer many opportunities to exploit forecasts in a cost-effective manner. But, on closer inspection, much of this potential is a mirage. Leaving aside the possible insuperable objection that the chaotic nature of atmospheric patterns in mid-latitudes will make worthwhile predictions unattainable, there is the issue of how much of the costs resulting from extreme weather is avoidable. In making a case for work on seasonal forecasting the US National Atmospheric and Oceanic Administration (NOAA) predicted[11] that 'a very broad range of users – and a substantial segment of the US economy – will benefit from interannual climate forecasts; the agriculture, fishing, forestry, energy, water resources, recreation, and construction sectors, to name several. In each of these, major losses that occur from ENSO events can be mitigated by prior knowledge of the event. For example, in the US agriculture sector alone, which was valued at some \$100 billion, it has been estimated that at least 10 per cent of the sector value is lost due to an ENSO event, but that some 15–20 per cent of that loss is preventable by a climate forecast that is right 3 out of 4 times. This suggests a loss-mitigation (savings) of \$4 billion due to ENSO forecasts over a 12–15 year period, for the US agriculture sector alone.'

How savings will be achieved is a matter for debate. As noted in the case of shorter term warnings nothing can be done about much of the damage caused by freezes, storms, droughts or floods. While increasingly accurate forecasts would lead to more cost-effective management of, say, fuel stocks and water reserves, only established success will influence more detailed planning decisions. Moreover, the complex response of many sectors of the economy will place additional demands on forecasters. For example, in the case of agriculture, as noted in Chapter 1, 1995 was a bumper year for British cereal farmers. Yet to most people the record-breaking heat and drought of July and August were seen as synonymous with global warming, and hence a threat to agriculture. Only accurate forecasts a year ahead of both the plentiful rainfall and mild conditions throughout the spring and the hot dry weather during harvesting could have told farmers they were not facing a repeat of 1976. Such accurate

and advanced predictions remain a long way over the horizon. So claims
that seasonal forecasts will soon lead to substantial reductions in the costs
of extreme weather may be a little far-fetched.

There is also a political dimension concerning public expectations of
these forecasts which is neatly encapsulated in a historical example.
During the unrelenting cold of early 1963 the British press made much
of the fact the US Weather Bureau successfully forecast the continuing
freeze. As a result of questions being asked in Parliament it was revealed
the UK Meteorological Office were preparing similar long range forecasts
on an experimental basis. In November 1963 the Secretary of State for
Air announced that monthly forecasts would henceforth be published.
These became the subject of much debate as their track record was, at
best, mixed. In 1980 their publication was discontinued as a cost-cutting
measure. So both the decision to start and to stop releasing these forecasts
had clear political overtones. In the meantime, the Meteorological Office
continue to do research longer term forecasts and to provide a service to
customers who are prepared to pay for it.

7.7 The statistics of climate change

The requirement to be able to predict several seasons ahead and provide
details of various meteorological variables elides neatly into the challenge
of providing economically valuable forecasts of climatic change. What
counts is not simply producing broad predictions of future warming, but
translating this into reliable estimates of the changing frequency of
extreme events. Once planners can estimate what the likely costs of such
events are and what investment is needed to cut these costs by a certain
amount then it will be possible to estimate whether such forecasting work
is good value for money. This question is only worth asking in terms of
future work. There is no purpose in asking whether past forecasting of
climatic change has been worth doing. At the simplest level it is a case of
bygones are bygones. In any case, the progress made on modelling the
climate has raised sufficient interesting issues to justify much of it on the
basis of curiosity research alone. But, having raised a series of major questions and underpinned international efforts to co-ordinate action to abate
carbon emissions, it now faces more searching scrutiny. This will probably
take the form of scientists being under greater pressure to set targets for
their work if they are to receive the funds they want.

The setting of targets will not sit easily with the great uncertainties associated with forecasting work. It will involve some accepted measure of skill which is attached to forecasts. If this is built round the development of ensemble techniques (see Section 6.2) it may be possible to establish measures of success which make forecasters accountable to those who fund their work. This could take the form of predicting a scale change over a given period together with an estimate of the likely scatter about the mean value. To take an example, it might consist of predicting that the average number of depressions each winter in a specific part of the North Atlantic with a central pressure of 950 mb or below will increase by $N \pm n$ per cent over the next 50 years. This would have the attraction of providing decision-makers with quantities for use both in making plans and in judging the value of the forecasts. If the increase (N) was large and the standard deviation (n) was small this would be not only of value to planners whose investments will be vulnerable to such major storms (e.g. offshore oil and gas operations in the northern North Sea and west of the Shetlands), but also capable of being evaluated relatively quickly. If, however, N was small and n big then the value of the forecast would be negligible and the prospects for early evaluation remote.

The complex nature of the models used to make predictions means this required degree of precision may not be achievable. But it needs to be estimated *in advance*. If the models cannot be tested, then the forecasters are not accountable to those who fund their work , and so it will be impossible to check whether increased investment in the forecasts is worthwhile. This may seem harsh, but the alternative is even less acceptable in terms of government spending. To be told in 10 or 20 years' time that not only is there still no obvious trend in the frequency of many extreme events, but also, in spite of investing in massive increases of computing power and ever more sophisticated models, we still do not know which of these events are likely to become more prevalent and, if so, by how much, just will not do.

So forecasters will have to become more accountable if they are to receive the funds they say they need. Monitoring of progress in terms of the economic benefits of modifying investment decisions to reflect predicted changes has to be part of this process. These estimates should include a wide range of planning and business investment decisions (e.g. building and zoning regulations to reduce the vulnerability of properties from storms and floods, managing the insurance market, agricultural practice, afforestation policy, sea defences, offshore safety). Without some measure of skill, which can be used to inform investment decisions and measure success or failure, forecasting work will become increasingly vulnerable to the 'business-as-usual' school of thinking, which is gaining support in the USA.

This frostier funding regime is not necessarily a bad thing for researchers into climatic change. It imposes a discipline which must be addressed if their work is to retain credibility. The broad range of the forecast scale of global warming for a doubling of CO_2 of 1.5 to 4.5 °C has not changed since the 1990 IPCC report, but the tendency to move towards the bottom end of the range has undermined the case for early action. Although the co-ordinated efforts of the climatological community resulted in the guarded consensus agreed by the IPCC (see Section 6.2) about the extent to which human activities are contributing to global warming, the forecasting of future economic impact has made little progress. The combination of downward revision of the pace of change and the absence of agreement on how regional patterns will shift in terms of both average conditions and anomalies must be confronted in making the case for additional funding. At the most basic level, the objective of immediate work must be to show whether or not we are able to forecast with any confidence that the frequency of, say, cold waves, drought, floods, heatwaves, hurricanes or winter storms will change appreciably in the next few decades. The value for policy formulation of being able to point out that using existing climatological statistics remains the wisest approach would be immense. Given the statistical minefield of very rare events, this scientific underpinning would, therefore, be worth considerable investment, rather than simply assuming business as usual is the right approach.

It follows that if the minimum objective is to reach conclusions on the unpredictability of extreme events, then more definite results on whether certain extremes will become more or less common will have greater value. So modellers will have to face the challenge of stating how well they think they can do, and living with the consequences. At the moment, various efforts to extract from GCMs some indication of how the incidence of extremes will change has produced rather limited results. If anything they show a tendency for the number of extremes to decline.[12] There is, however, a feeling that this may reflect the limitations of GCMs to simulate accurately major factors in climatic variability such as the ENSO. So, without improved models this crucial question will remain unanswered.

This challenge of predicting the importance of future extremes is part of a wider issue of seeking to find optimum solution for multivariate problems.

Even in quite simple examples it can be shown that finding an optimum solution is impossible. So for something as complicated as estimating the optimum economic response to possible changes in the climate is beyond us. Instead we should accept that irrespective of where we wish to get to, we have to start from where we are, not where we would like to be. Given the immeasurably large number of outcomes arising out of exploring the full range of options available, we have to rely on a process which is often referred to as 'local optimisation'. This explores in a random manner using, say, ensemble techniques the possible outcomes when we move a small distance from where we are now, while allowing other features of the world around us to change as well. If the result of modelling our local phase space is to reveal a nearby set of circumstances which look appreciably better than continuing as we are then this should influence policy. But, if there does not seem to be scope for real improvement by incremental movement in a given direction, then we need to be very careful about being stampeded in any specific direction because of a single perceived threat.

This means that in the real world, forecasts of climatic change not only have to be frank about what they can and cannot predict but also must be viewed in the context of the other challenges facing societies. Where the response to a climatic threat is consistent with sustainable and economically attractive solutions to the other environmental and social challenges we face then the case for action is correspondingly greater. But, where the options for dealing with different challenges are contradictory, then the arguments for sitting on our hands are that much stronger.

7.8 The balance sheet

Having given forecasts a pretty rough time, we must come back to the basic question of whether they are really good value for money. This boils down to asking whether too much or too little effort is being devoted to this work. The other side of this question is whether too much or too little attention is paid to forecasts. Clearly in the USA many Republican politicians take the view that too much is being spent on climatic change forecasting and too much attention is paid to the results of this work. On the other hand many environmental groups would retort that, while much of the work needing to be done is in progress, far too little is being done to heed the warnings. So, talk of cutting back of research is nonsense, and what really matters is for governments to get on with honouring their guarded commitment to reduce carbon emissions agreed at the Berlin Convention in 1995.

Clearly, there will be no meeting of minds. What will decide the balance between the competing arguments is weather events over the coming years. In the meantime, forecasts will need to build on the progress they have made so far. This requires them to target their output to reach their perceived audience or to meet the needs of specific customers. In the case of numerical weather forecasts up to 10 days ahead, this will involve forecasters ensuring that their predictions are accessible to as many users as possible. In addition they will have to set tough and quantifiable performance targets so that they can demonstrate the real progress to justify investment in more and more powerful and expensive computer systems. Otherwise they will have to come to terms with having to manage with slightly more modest expectations of support for their work. This may not be an altogether bad thing as it may lead to going back to basics to unravel some of those puzzles of atmospheric physics which may hold more promise of improving certain aspects of forecasts than relying on ever heavier number-crunching. In particular, improved treatment of clouds, fronts and explosive cyclogenesis, and physics of how the atmosphere if affected by the changing characteristics of the land surface could all help produce better short-term forecasts. On the slightly longer term, getting a better fix on what triggers the switch between quasi-stable weather regimes could secure substantial advances. Furthermore, demonstrating that the improved treatment of all these processes leads to better weather forecasts will also bolster confidence if they are incorporated in GCMs to produce better predictions of climate change.

Seasonal forecasts and predictions of climatic change will, nevertheless, face a stiffer test. There is no proven track record. But unless an accepted measure of success can be generated soon, this work runs the risk of coming to be regarded as a fascinating but esoteric scientific pursuit: forecasters cannot adopt Humpty Dumpty's approach and believe that success means just what they choose it to mean. They must produce results so that the application of forecasts output contributes measurably to the solution of problems. In this respect, if their work also helps find manageable and realistic solutions to other environmental problems (e.g. deforestation and soil erosion), it is more likely to receive recognition and further funding. At the moment, however, past forecasting work appears on the balance sheet as an asset, but unless it starts generating more tangible benefits soon, future additional value will be called more and more into question.

7.9 Notes

1 Rosenzweig (1994).

2 Hulme *et al.* (1992).

3 Krishna Kumar, Soman & Rupa Kuma (1995).

4 Kripalani & Kulkarni (1997).

5 IPCC (1995), p. 337.

6 Broecker (1995b).

7 Gray (1990).

8 Landsea *et al.* (1994).

9 Colorado State University press release, 5 April 1996.

10 Chang, Ji & Li (1997).

11 NOAA press release, November 1995.

12 IPCC (1995), p. 336.

8

Conclusions

'Now, *here*, you see, it takes all the running *you* can do,
you must run at least twice as fast as that!'
The Red Queen, in *Through the Looking-Glass*, Chapter 2

So does the weather really matter? Clearly, extreme weather events and
longer term climatic change have both exerted substantial influences on
economic and social history and played a fateful role in other aspects of
history. But what has not been established is that they have altered the
mainsprings of history. These have been energised by the vast forces of
population growth, political and cultural developments, industrial advance
and scientific and technological progress. But, whenever the social structures have become overextended, adverse weather has often been the catalyst of breakdown and failure.

As for things getting worse, there is little evidence that the weather is
becoming more extreme. In spite of Tom Karl's[1] analysis that certain
types of extremes in the USA, which might be symptomatic of the greenhouse effect, appear to have become more prevalent since the mid-1970s
(see Section 5.4), the overall picture presented both in the many time
series in this book and in the 1995 IPCC report[2] is of no significant
change.

If correct, these low-key conclusions lead naturally to the question –
what is all the fuss about? Is it simply that in a more media-orientated
society we are being led to believe that every major snowstorm is the
'Blizzard of the Century' and every drought is the end of civilisation as
we know it, or are there more profound issues to address? Whatever the
temptation to embellish current events, it would be foolish to dismiss it
all as media hype. The assumption that as a result of human activities the

climate is sailing into uncharted waters is bound to have a profound impact on our thinking, even if more extreme weather events are not on the cards yet. What really matters is whether our social structures are becoming more, or less, vulnerable to inclement weather.

8.1 So far, so good?

Looking at where we have got to in recent years, the balance sheet suggests that, if anything, we have become less vulnerable to weather fluctuations. This improvement has, however, been swallowed up in rising expectations. So, while death and suffering have been dramatically reduced in the developed world, and there is substantial scope for transferring these benefits to the developing world, in the wider economic context the costs of weather damage have risen dramatically. This implies that we now expect to be fully compensated for loss, damage and inconvenience while being able to conduct our normal day-to-day business, which puts more and more property and possessions at risk and involves more and more activities which are vulnerable to weather.

This assertion is best examined in terms of citing practical examples of how rising expectations involve disproportionate costs, before making general observations about how these may be self-limiting. Starting with the examples of cold winters discussed in Chapter 3, we see that while many aspects of society, such as energy supplies, have become more flexible, the disruption has transferred to many individual activities. So, even when accurately forecast, major snowstorms often cause chaos because many people expect public services to keep roads open and maintain transport systems. Accordingly they put their vehicles and commercial enterprises at risk on the assumption that their insurance will cover any loss.

Although property losses cannot be treated in quite so cavalier a manner, the expectations of many people that the risks of developing and living in coastal areas which are vulnerable to winter depressions or tropical storms will be absorbed by the insurance industry at reasonable cost have to be kept under constant review. In the same vein, the standard image of hurricane damage showing heaps of stunningly expensive yachts and powerboats piled up in the corner of some marina (Fig. 8.1) suggests their owners will have to come to terms with the real risks involved in investing in these playthings. At the more mundane level, homeowners will have to be disabused of the assumption that gained some popular

Figure 8.1. Damage caused by Hurricane Andrew when it hit South Florida
in August 1992. (Reproduced by permission of Popperfoto.)

currency in Britain, following the storms of October 1987 and January
1990, that spending money on roof repairs and maintenance was pointless:
all it needs was a good gale and the insurance companies would buy a
new one, as after major disasters claims were not scrutinised closely. So
the insurance industry will have to do more to check such claims in the
same way as isolated accidents. It will also have to do more to make
certain that local authorities apply building codes and zoning regulations
effectively to reduce the vulnerability of property to storm damage.

In principle the process by which insurance premiums are adjusted to
reflect the latest perceptions of risk will cover the cost of the rising vulnerability and in due course may alter people's behaviour and eventually curb
the cost of risky behaviour. But the process is slow and can be dislocated
by two major forces. First, a series of major disasters, whether or not due
to climatic change, can in theory overwhelm the insurance market. Events
in the late 1980s and early 1990s showed that this is a real threat. So it is
possible that a sudden upsurge in massive weather disasters could leave
the insurance industry unable to meet the needs of the market. Even if
this apocalyptic combination never happens, the prospect of a rising but
unpredictable trend represents a powerful threat as future losses are paid
out of reserves built up from past premiums. Too rapid a rise in the

number of disasters makes it impossible to set premiums at economic levels and the creation of some government-backed fund may be the only option. So, both the insurance market and government have an immense interest in getting improved forecasts of the future incidence of extreme weather events.

Hurricanes are the clearest example of this problem, although other severe storms in the USA (e.g. winter snowstorms and tornadoes) are also a major threat to the insurance industry. Since we do not know whether Hurricane Andrew or the stormy season of 1995 were one-offs or the start of a new trend, the insurance industry cannot distribute the risk in an informed manner. In the meantime, it must plan on the assumption that in the next few years an intense hurricane could hit a major urban area on the US Gulf Coast or East Coast causing $50 billion in insured losses. If this came shortly before or after a major earthquake in the USA then the load on the insurance industry might be overwhelming.

The second problem is the related issue of a sea-change in the types of losses incurred. A good example of this phenomenon was the damage due to subsidence during the summer of 1976 in England. This type of cover had only been introduced into domestic policies in 1971. So, the arrival of such an extraordinarily hot, dry summer so soon after this change provided a stern test for the insurance market. The costs in 1976 plus the consequent changes in building regulations led to a fundamental reappraisal of domestic property insurance which could not have been anticipated before the long hot summer. The fact is that the hot dry summers in 1983 and 1984, and again in 1989 and 1990 have kept the pressure on the industry. The cost since 1976 has reached around £3 billion ($4.8 billion), with peak pay-outs of some £500 million ($800 million) in both 1990 and 1991. As a consequence insurance rates per unit cover had risen 50 per cent by the early 1990s. But the relatively modest increase of some £165 million ($260 million) after the drought of 1995, together with competitive cuts in the rates of a few per cent, suggests that this type of damage had become an integral part of the domestic property insurance. It is, however, a measure of how deeply the subject had entered the British psyche that fear of subsidence was blamed for the faltering recovery of the housing market (see Section 6.4) in late 1995.

So, the evidence suggests that the insurance market remains the most efficient means of achieving the right balance for covering the risk of future weather damage. To the extent that certain sectors become more vulnerable they will have to pay higher premiums, which will be reflected in the desirability and economic value of businesses and properties in these risky sectors. In this way, where expectations incur too high a risk, individuals will have to learn they cannot make the assumption that the rest of society will share their burden through the insurance market. Alternatively, the insurers may look at ways of sharing the burden with those seeking insurance by only covering a proportion of the loss, or only covering losses above some figure. If the losses are distributed between the insurer and those incurring the losses then there are powerful forces operating to keep them as low as possible. Here again the role of government, both local and national, in regulating developments and enforcing regulations is an essential part of maintaining an orderly market. In the last resort, insurance companies can, of course, choose not to provide cover for certain forms of risk. This will, however, mean that major disasters end up in the lap of government, one way or the other.

Alongside changing people's expectations there is the fact that certain weather events will alter fundamentally their perceptions of risk. If the substantial development of the US coastline of the Gulf of Mexico was, in part, stimulated by the perceived low risk of hurricanes resulting from their declining incidence and intensity in recent decades (see Fig. 1.1), the record-breaking season of 1995 changed many minds. As Hurricane Opal became the fifth storm to hit the coast in four months, many inhabitants were seriously questioning whether it was a sensible place to live. Their new insurance premiums will no doubt reinforce their sense of unease. This potential reaction is reflected in a recent analysis of the economic costs of greenhouse-induced sea-level rises for developed property in the USA.[3] This considers a range of sea-level rises from 0.3 to 1.0 m by the end of the next century. Depending on whether or not people exercise foresight, the costs could reach $100 million to $1 billion per annum by the year 2100. These figures are about an order of magnitude lower than earlier estimates and reflect the inclusion of the cost-reduction potential of natural market-based adaptation in anticipation of the threat of rising seas.

8.2 Choppier waters ahead?

Future vulnerability to whatever weather and climate fluctuations are in store for us will depend on whether the needs and expectations of a growing population can be met by an expansion in agricultural and industrial production while maintaining an acceptable level of environmental quality. The more the achievement of these objectives depends on the overexploitation of marginal resources or the adoption of environmentally damaging practices, the greater the prospect that the weather will loom larger in the problems we will have to confront. There is no agreement as to whether or not these challenges will be manageable. At a time of huge economic and political change throughout the world, trying to isolate the role that might be played by future climatic change may, in any case, be an oversimplification. If, however, we accept this narrow perspective, then at the simplest level the answer depends on whether you are an optimist or a pessimist. The former would argue that solutions will be found within the context of vigorous but sustainable economic growth. The latter would take the view that current growth is unsustainable and without radical adjustments to economic policy it will all end in tears.

Which camp proves to be right depends on two things. First, whether the climate experiences a gradual shift to a warmer world or undergoes larger, erratic and irreversible shifts to a new global order – an issue on which computer models cannot yet provide reliable guidance. Second, whether global economic, political and social structures are capable of adapting to the predicted changes let alone more unpredictable fluctuations. Broadly speaking the optimists are more likely to be proven right if the more modest forecasts of warming are correct. In these circumstances, the global economic system may be sufficiently adaptable – especially if the political resolve to reduce the future build-up of greenhouse gases in the atmosphere is sustained at the expense of some economic growth. The greater the uncertainty that these conditions can be met, the greater the likelihood that the pessimists will be vindicated. In particular, the combination of rising population pressure and environmental damage, such as loss of topsoil and deforestation, which could amplify the impact of climatic change, are cause for grave concern.

Looking first on the bright side, everything depends on developing confidence in both the forecasts of global warming and our ability to monitor the true nature of climatic change. The earlier it is possible to reinforce the conclusions of the 1995 IPCC Report (see Section 6.2) with a more categoric assertion that what we are seeing cannot possibly be due to the natural variability of the climate and is consistent with the predicted impact of human activities, the easier it will be to sustain economic and political action to cut back on the most damaging activities. This mutual reinforcement of measurements and models depends on improvements in

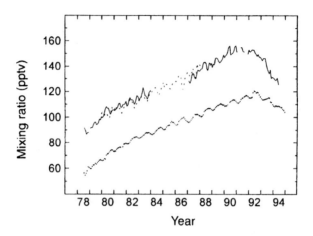

Figure 8.2. The impact of the international agreements to curb the emissions
of chlorofluorocarbons (CFCs) can be seen in the decline in methyl chloro-
form levels in the atmosphere since around 1991 (records from Ireland,
Oregon and Tasmania). (From IPCC, 1995, Figure 2.10(b).)

both areas. In the case of getting the most precise picture of climatic
change it is a matter of ensuring that current measurements are used with
the utmost effectiveness and any that untapped source of information on
past climatic change is fully explored to provide improved insight about
natural variability.

The underlying message here is that action flows from being confronted
by overwhelming evidence of a real and common threat. The tortuous
progress of international action on climate change contrasts with the rapid
action taken on cutting emissions of chlorofluorocarbons (CFCs) following
the discovery of the 'ozone hole' in the mid-1980s (Fig. 8.2). While the
scale of action required to have a real impact on the emission of greenhouse gases is much greater than cutting back on CFCs, the level of
uncertainty is the crucial factor in holding things back. Anyone who has
been involved in difficult international negotiations with many countries
knows full well the stately minuet of refining agreed documents, which
can grind to a complete halt where the conflict of interests of different
groupings leads to doubts about the reality of any threats and uncertainties
about the effectiveness of any proposed action. Once bogged down in the
mire of doubt, the whole process can descend into an exercise of drafting

ever more inward texts which give the impression of a grand design while enabling individual countries to carry on doing what they had intended to do all along. While this process of negotiation is probably better than not talking at all, it is dreadfully slow and wasteful.

So, whether at the international or national level, the urgency in developing greater confidence in the causes of current fluctuations cannot be overstated. The political attractions for governments of putting off expensive and unpopular action are powerful. The resistance to action such as raising the federal tax on gasoline in the USA or putting value added tax (VAT) on fuel in the UK shows just how hard it is to introduce fiscal measures which might reduce CO_2 emissions. Furthermore, the feverish attempts by some representatives of the energy industries to identify an environmental 'conspiracy' in producing the final conclusions of the 1995 IPCC report, show it is not possible to separate scientific analysis from political realities. So scientists, who are closely involved in the prediction of climatic change, must recognise that they are working in an area where their competence will be challenged and their motives questioned by those whose interests are threatened by action to prevent global warming.

Improved explanations of variability are needed to underpin the scientific analysis and to enable governments to handle the arguments of vested interests who will exploit the normal ups and downs in the weather to claim that action is premature. Because, as the many examples cited in earlier chapters have shown, the interannual variance of the seasons is far greater than any of the trends so far identified in climatological statistics at a regional level. But action will have to be taken by governments on the basis of avoiding damaging consequences in the shift in the balance of extreme weather events. So in order to make sensible decisions there is a pressing need to understand what is causing change and then to predict how the incidence of extremes will alter in the future.

Nailing down the causes of variability is equally important to those already convinced of the need to take action to prevent global warming. The proper interpretation of the meaning of any single extreme event, or a run of events, is essential. The temptation to rush to judgement when public interest in the weather is heightened is not helpful, if, on mature reflection, it has to be accepted that what was observed was not much out of the ordinary. There have been too many examples of extremes being seen at the time as harbingers of predicted change, only to subside back into the noise when viewed in the light of subsequent ups and downs. If we really are now facing up to changes whose costs would far exceed those of prudent early prevention, then picking out the events which confirm this fact is what really matters to those who are pressing for action.

The first priority is to corroborate the predictions of the GCMs by establishing the precise nature of the current global warming. This must discriminate between what can reasonably be defined as natural variability, notably due to longer term fluctuations in the oceans, volcanoes, and possibly, solar activity, and what can only be put down to human activities. This requires an unambiguous measure of temperatures across the globe, both at the surface and at various levels up through the atmosphere (the 'fingerprint' of human activities – see Section 6.2). In spite of the IPCC concluding that current temperature trends are hard to explain in terms of natural variability, there remains the nagging doubt that we have not got to the bottom of what is causing current changes. The first stage must, therefore, be to resolve the uncertainties in available measurements of past and present trends, both globally and regionally, and to establish whether this increases our confidence in the conclusion reached by the IPCC in 1995.

This work will need to explain why certain places have seen changes which are regarded as consistent with global warming, whereas others have remained unaffected or have even experienced apparently contradictory shifts. Many of the examples cited in earlier chapters show no appreciable trends, and this is bound to influence governments facing difficult funding decisions. The models will have to come up with much more convincing evidence of why, even though some important features of the climate have remained constant (e.g. storminess in north-west Europe and Indian monsoon) or even declined (e.g. the incidence of North Atlantic hurricanes), climatic change is the concern of all nations and not just those threatened by current events. As the temporary surge in grain prices in April 1996 demonstrated (see Section 4.5), it can still be a good old-fashioned cold winter and late spring which generates significant market activity. So, getting to the bottom of these conflicting signals will help to identify where our economic and social structures are becoming more vulnerable to current extremes.

This understanding will also be an essential component of developing concerted international action on carbon emissions. The fact that so much of the current emissions comes from the advanced industrial nations, whereas future growth will come from developing countries, is the source of major tensions. Because modest reductions in emissions in North America or western Europe, if transferred to countries like India, would enable them to continue to achieve rapid growth, these issues have to be addressed. Only if there is agreement on what the impact of climate change is on different parts of the world can there be any prospect of negotiating reductions in emissions which reflect both the best approach to reducing damaging shifts in the climate and the economic needs of developing nations. Without this agreement there is a real risk that the negotiations will be polarised, with the developing nations concluding that the entire global warming thesis is a conspiracy designed to protect the interests of the advanced industrial nations and prevent the others from making progress to reduce the gap between rich and poor.

The challenge of the regional variations in global warming assumes even greater significance when we turn to extreme weather events. Because the most vulnerable – be it sectors of society or nation states – suffer most in weather disasters, any reliable emerging conclusions on shifting patterns of the incidence and geographical distribution of extremes will play an even greater part in future international negotiations. So the translation of broad global trends of climate change into predictions of what this will mean for the incidence of extreme weather events around the world is vital. A mean warming of a degree or two does not strike fear into planners' hearts. It is more frequent droughts, heatwaves, floods, storms and even cold spells which will do the economic damage and so justify the case for early prevention. If a gradual warming does no more than lead to slow shifts in the frequency of these extremes then the case for radical action is undermined. The priority for modellers is to be able to show whether or not the non-linear response of the climate to any warming is a more malign prospect and if so who are the big losers. If they can do this to the satisfaction of politicians and business planners then the case for action will be strengthened. If they cannot, they will face increasing difficulty in sustaining the existing fragile international resolve to curb the emissions of greenhouse gases in the twenty-first century.

What is being asked of modellers may be impossible. The unpredictability of non-linear processes could mean that GCMs cannot produce convincing results about the shift in the incidence of extreme weather events. Predicting shifts in the frequency of winter regimes in the northern hemisphere is beyond current GCMs (see Section 6.2). Such changes are, however, central to any estimates of the economic impact of climatic change. Similarly a measure of how the summer monsoon over the Indian subcontinent and south-east Asia will shift may be unattainable. Without quantifiable forecasts countries in this region have good reason to balk at economic sacrifices to avoid worsening climatic disasters. An authoritative view on how the incidence of extremes should change is the only way to avoid a sterile debate between opposite camps. Each new disaster will be greeted by a polarised response. Either it will be seen as nothing out of the ordinary, providing a sufficiently lengthy view is taken, or it will be heralded as the latest example of the inevitable consequences of global warming. For the modellers, continued sitting on the fence over the question of extremes will prove increasingly uncomfortable, however intellectually impeccable it may seem.

8.3 In praise of gradualism

When surrounded by seemingly impenetrable thickets of uncertainty, crashing off into the undergrowth in search of a way out is a risky option. A gradual approach has many attractions and is supported by the analysis of economic models (see Section 6.5). This does not mean doing nothing. At the practical level, a suitably balanced, multidisciplinary attack on all aspects of global warming must use the widest set of criteria to set priorities to reduce uncertainty. There are three main prongs to this attack. First, there is the basic issue of improved research to underpin the efforts to fingerprint current and future changes in climate. This is not just a matter of addressing the questions of the real impact of the wide range of human activities on the climate that have already been identified, but how these are linked to other pressing ecological and environmental issues. Where the combined effects of these impacts extend well beyond simply the climatic impact then the economic case for radical action is stronger.

A good example of this combination relates to urban environmental problems. Half the world's population lives in urban areas and the proportion is rising. Many health problems of global warming are amplified by the effects of the urban heat island (see Section 6.5), especially in the big cities of warmer temperate regions (e.g. New York, Shanghai or Tokyo). The consequences of more frequent and intense heatwaves will be compounded by other forms of pollution. So pressures to reduce vehicle emissions, promote clean public transport and reduce power consumption related to air-conditioning will reinforce the case for action in cutting down the release of greenhouse gases. In the case of cutting back vehicle emissions and increased efficiency of energy use, any action will directly reduce the release of greenhouse gases. But indirect strategies, such as planting more trees to create shade in suburban areas and reducing the absorptivity of the surfaces of buildings, parking lots and roads may be an even more cost-effective means of preventing the consequences of global warming becoming intolerable.

Similar economic arguments apply to improved management of both tropical and boreal forests (see Section 6.2), prevention of desertification and salinisation of semi-arid regions and reduction of soil erosion by all forms of agriculture. While afforestation is only a temporary solution to reducing the build-up of CO_2 in the atmosphere, if it has other climatically beneficial effects, the economic case for its wider adoption will be correspondingly strengthened. Only by developing a balanced approach to these issues can we hope to find solutions to the challenges confronting us which recognise and give appropriate weight to the competing arguments for different actions. Furthermore, this balanced approach, which addresses many of the other pressing issues confronting the developing world, is more likely to be seen as part of a constructive strategy rather than a western conspiracy to maintain or increase existing inequalities.

This holistic objective applies equally to the second prong of an incremental offensive. This is the question of the development and exploitation of new technology. Because many of the economically attractive options for reducing the impact of global warming will depend on new technologies, the timing of their exploitation is crucial. Large-scale investment ahead of real need leads to substantial misallocation of resources. This is of particular relevance in energy generation systems where investment decisions affect the mix of production for decades ahead. The over-reaction in the UK to the cold winter of 1963 is a good example of how long it takes energy investment decisions to work through the system (see Fig. 3.4).

At all levels difficult decisions will have to be made about energy generation systems in order to balance economic and political demands. Future investment in major generating systems, notably power stations, will have to reflect the resolution of current debates on the acceptability of alternatives. At present nuclear power is widely regarded as even less acceptable than burning fossil fuels. But, without substantial changes in consumption patterns this may be the only way of meeting a significant part of future demand. The environmental and economic arguments about many renewable options are still in their infancy. The vehement opposition to any further increase in hydroelectric power is a measure of what may lie in store for other renewables like wind turbines, which are economically competitive in some circumstances. At the same time, new sources of energy must be allowed to compete on a level playing field. Where the production of fossil fuels is subsidised or receives tax breaks this support needs to be phased out. What is certain is that politicians will have to establish difficult compromises if their electorates are not to end up freezing in the dark: they surely won't wish to find themselves in the position of Emanuel Shinwell in 1947 (see Section 3.2) or Jimmy Carter in 1977 (see Section 3.5).

On a smaller scale awkward questions also lie ahead for all of us. Dewy-eyed appeals to adopt solar power or electric vehicles can do more harm than good. While passive solar heating of housing is already a sensible design feature for new buildings, photovoltaic solar panels have a steep economic hill to climb. If they are going to be a paying proposition it will be in Arizona or other sun-drenched parts of the world that they first make the breakthrough. So let the citizens of Phoenix lead the way – those of us in Boston or London should sit on our hands for the time being. As for electric vehicles, the issue is getting the analysis right in terms of both the economics, environmental impact and balanced transport policy. Airily talking about 'zero-emission vehicles' without taking proper account of the emissions of the electricity-generating plant, the environmental impact of increased transmission systems and the pollution controls needed to supply, service and dispose of the batteries with all the nasty chemicals involved, must not be underestimated. Leaving aside how these issues may merely reinforce the case for improved public transport, including electric vehicles in cities, it suggests this is an area where economics should rule. If with proper controls, electric vehicles can gain a foothold in the market, so be it. If, however, they require special support, then their case should be subject to more searching scrutiny before the economic scales are tipped in their favour.

By way of contrast, many of the choices for reducing energy consumption and improving the efficiency of energy use are not only already available, but also are economically attractive. For both the developed and developing world, ways must be found to remove institutional barriers to the swift and effective exploitation of these technologies. At the same time a continual process of technology assessment has to be an integral part of the evolving strategy to handle the emerging view of what are the most important threats associated with future climatic change. Furthermore, where options look particularly attractive, efficient dissemination of information to spread good news is vital. In particular, the effective use of information technology to enable the developing world to share in appropriate technologies quickly is in everyone's interest.

The third prong of the attack has to be the development of a political climate in which fiscal measures can be taken to reduce the emission of greenhouse gases. If introduced suddenly, high taxes on fossil fuel consumption or carbon emissions will be unpopular with the electorate, especially if it hits the poor and elderly hardest. But, as part of a progressive shift in how governments raise revenue, and with the stated strategic objective of minimising future climatic change, they may come to be more acceptable. Just as taxes on tobacco, alcohol and other forms of consumption are seen as reasonable means of raising revenue, so taxes to prevent global warming can become accepted. This shift does, however, require the support of all political parties otherwise it will become a political football which is booted around for short-term electoral advantage. In these circumstances it would become more, rather than less, difficult to introduce substantial fiscal measures to reduce carbon emissions. So any move in this direction must be part of a sustained shift in how revenues are raised and be substantiated by growing evidence that action is needed to prevent greater economic losses. All of which can only be part of a gradual but coherent, multidisciplinary strategy to reduce the costs of the certain threat of global warming.

Another facet of the political climate is handling the range of reactions to changing events of both individuals and institutions. These cover the gamut of the perverse defensive responses to perceived threats through to reassuring adaptability learnt from experience. So developing awareness which prevents the equivalent of buying electric fires in the hope of avoiding coal shortages (see Section 3.2) or overwatering lawns at the first hint of drought will pay substantial dividends in maintaining a balanced response to a shifting frequency in weather extremes. At the same time authorities will have to seek to moderate consumers' expectations, if the cost of preserving current levels of security of supply becomes too high in a more variable climate. Conversely, convincing drivers they cannot have untrammelled access to urban areas, especially during heatwaves, and that they will have to pay much more for the privilege, must be a long term goal. Alongside these processes, supporting and developing adaptable responses to potential disaster due to, say, flooding or freezing conditions will both reduce the immediate costs and promote longer term strategies to enable communities to live with climatic change. This evolutionary approach is also the only way in which policy can adapt to sharp swings in public attitudes to the economic consequences of cutting back on greenhouse gas emissions.

Developing greater coherence in dealing with the threat of climatic change will also make it easier to decide when direct action by governments is called for in responding to weather disasters. This could involve support of communities both to enable them to re-establish themselves, or to enable them to move to a safer environment. The response of the Roosevelt administration to the Dust Bowl years is a good example of the first type of reaction. The fact that farming communities in the Midwest have largely survived the ups and downs of the climate since the 1930s confirms government intervention was sound. Even more interventionist is preventing the re-establishment of communities which have become too vulnerable to certain forms of weather disaster. The action taken after the Mississippi floods of 1993 is a good example of this process. If done in the right way it not only makes economic sense but also creates a better political climate for taking courageous decisions in the light of any emerging evidence of unacceptable climatic risks of a business-as-usual strategy. This sense of political realism will become an increasingly valuable commodity if we are facing a more uncertain climatic future, so governments should seize the opportunity to confront the electorate with the economic facts of life.

Having explored the more optimistic case in some detail we must return briefly to the pessimistic view. I say 'briefly', not because it is necessarily wrong, but because its implications can be stated bluntly. If irreversible and substantial changes in the climate are already in train, it is probably too late to introduce the radical institutional changes needed to make the major adjustments to the threats facing us. Furthermore, in such circumstances meeting the challenge of climatic change will be only a subset of wider environmental and social problems facing both national and international institutions. So it follows that these too will only be confronted after a great deal of pain and conflict. In short, if a climatic Armageddon is upon us, then the only solution will be to find ways of living through it as it unfolds, because the political fact of the matter is that adopting painful remedies will only be accepted when it is clear, beyond peradventure, there is no alternative.

8.4 Taking the longer term view

If the optimistic view that a 2 to 3 °C rise in mean global temperature by the end of the twenty-first century is something that we can take in our stride, is there any other reason to worry? Two additional thoughts are worth considering. First, there is the possibility that human activities could trigger a far more erratic response and plunge the world into some fundamentally different climatic regime. Whether this was the result of the incidence of normal weather patterns being fundamentally altered by the warmer conditions, or driven by some non-linear switch in the circulation of the oceans, it could result in many parts of the world experiencing much greater and more sudden changes than implied by the standard predictions. But, by their nature, such chaotic responses to the perturbations caused by human activities are unpredictable. This means that, while the most obvious candidate for triggering this sudden change is the build-up of greenhouse gases, any substantial perturbation, whether due to human activities and/or natural fluctuations, could set things in motion. So, short of recommending all such activities should cease, we have to hope and pray that the climate is sufficiently stable to suppress any over-dramatic response to change. If not, then all we can do is hang on to our seats as it will be a bumpy ride.

The second point is a more philosophical one. This is that we are taking too short term a view. In any normal circumstances, suggesting that looking only as far as the year 2100 is myopic would seem ludicrous. Not only is this far beyond any normal planning horizon, but also, given the conclusions reached on the inability to forecast well ahead, it seems unwise to attach any significance to current thinking about what may happen in the twenty-second century. There is, however, one reason to question this conclusion. If we believe the global climate will warm, say, 2.5 °C by 2100, it follows in the subsequent decades this warming will continue. At some stage, depending on whether the warming rate remains steady or accelerates, temperature levels, and rises in sea levels will become intolerable. Among other things, the ice sheets of Antarctica and Greenland will eventually collapse, inundating huge areas of low-lying land around the world. This alone would justify substantial preventive action. So if this warming is inevitable unless action is taken, the question becomes when is the best time to start investing in prevention of distant but unacceptable changes?

The key issue in making decisions, having established what the costs of prevention are and what the likely benefits will be, is what discount rate should be applied to the future stream of benefits. For most economic decisions the discount rate is linked to interest rates on low-risk stocks and bonds. Typically a figure of 5 to 10 per cent per annum is used, depending on the level of risk involved in the investment. It is also accepted that public sector investments can use a slightly lower figure because risks are spread over a large number of taxpayers, rather than a smaller number of shareholders, and also because they displace private consumption. But this differential cannot be too large or public investment will displace more productive private investment, and also it must be applied consistently to public sector investments, otherwise one area will swallow up a disproportionate share of tax revenues. So, while a figure at the lower end of the standard range, say, 5 per cent, can be justified, much lower figures require special pleading. As for the longer term effects of global warming, a figure of 5 per cent is a killer. Any benefits in the year 2100 and beyond accruing from an investment now would have to pay a return of more than a hundredfold in real terms. In an uncertain world, there can be no economic argument for making such an investment. The only way to justify investing now for something so far in the future is by using a discount rate of close to zero per cent.

In his book on the economics of global warming, which I drew on heavily in Chapter 6 in discussing modelling the costs of reducing carbon emissions, William Cline argues fervently for a special low discount rate for preventing the build-up of CO_2 in the atmosphere.[4] One way of looking at the emotional appeal of what is often termed 'intergenerational justice' is to consider, as an economic issue, the example of why we plant trees. On the face of it, if the only benefit of growing trees was the timber they produced, the application of a 5 per cent discount means we should not bother to plant many of them. This is because the other benefits, such as woodland management for recreation, are far less amenable to economic assessment. Indeed, during the 1960s and 1970s the UK Forestry Commission convinced the Treasury that their investment decisions should be subject to a much lower discount rate of 2 per cent. To the extent that the Commission was still fulfilling its original strategic aim, for which it was established during the First World War, of providing a secure supply of timber for pit-props in the coalmines, this would have been a sensible decision. In practice, this role had long since ceased and what the decision implicitly recognised was the wider amenity value associated with managing large areas of the nation's forests. This acceptance that we do not plant trees solely for their timber, but because during their lifetime they serve a number of aesthetic and social functions for successive generations which have economic value, justifies not imposing a standard discount rate. Indeed, many trees are planted with no thought of their economic output, and reflect that, for many of us, a world without plenty of trees is inconceivable.

The relevance to global warming of this is, apart from the fact that afforestation has the benefit of locking up considerable quantities of carbon, to put William Cline's arguments in a simplified form – a world with an average temperature 10 °C or more warmer than the present is inconceivable. But it does not follow that a low discount rate must be applied immediately to any investments designed now to prevent this unbearable prospect. Leaving aside the fundamental problem of deciding which investments would qualify for the special financial regime, there are two obvious reasons for caution. First, there is a considerable risk that the additional investments may not achieve their intended objective. Second, there are inevitable political pressures over the distribution of public sector investments as carbon reduction initiatives use money which might otherwise have been spent on, say, health services. So, as with growing trees, we have to be certain that what we invest in will deliver precisely what we want.

When considered alongside the competing demands for investment in transport, social, health and education services, it is hard to see how a special case can be made for diverting resources into preventing warming beyond the year 2100. The appealing philosophical argument about how much we should pay to preserve the future environment for our grand-daughter's grandson can be applied with equal force to the benefits of maintaining current essential services in our society because they will contribute as much to the well-being of future generations. It is unlikely that governments will be persuaded to divert scarce resources to such altruistic ends – they will win more votes on concentrating on the here and now. So political realism points to the incremental approach being the only option. But if responsive to the longer term it is capable of making investments now which can make substantial contributions to the challenge of climatic change over the years to come.

The question of far-sightedness also arises in the matter of preventing the longer term ecological impact of the pace of global warming (see Section 5.1), but requires us to look at things in a different light. This revolves around the fact that the predicted warming could be far more rapid than anything experienced in the last 10 000 years. This is true, but on an evolutionary timescale this period of climatic stability is but a blink of the eye. So, leaving aside the other aspects of human activities, which may lead to far more rapid destruction of habitat, we can only consider the ecological impact of future climatic changes by taking a longer look backwards at the far more dramatic changes that occurred during and between the ice ages (see Fig. 5.2). In this context, a good starting point is the observations by Richard Dawkins that the animals and plants we are concerned about are probably out of date because they were built under the influence of genes that were selected in some earlier era when conditions were different.[5]

The evolutionary perspective suggests that for most species their genetic make-up has not caught up with the benign climate of the last 10 000 years. This means they retain the defences which enabled them to adapt to the more chaotic climate of the previous million years or more. If human activities lead to a rapid warming, and possibly to a more variable climate, flora and fauna are likely to get by rather better than the social structures we have created and which are so finely tuned to current climatic circumstances. The fact that we are also likely to have the basic genetic armoury to adapt to the consequences of greater climatic change is scant consolation. In terms of preparing for the economic and social challenges of the future, however, the gradualist approach remains the most politically realistic option.

8.5 Competing for attention

All of this suggests that those working on the issues associated with climatic change and global warming will live in the type of world described by the Red Queen – at the current rate of progress they will be lucky to stay in the same place: to get more of the funding cake they will have to run at least twice as fast. This is because we have not identified clear evidence that climatic change represents a proportionately greater threat to our economic well-being in the future than in the past. So, to justify more attention this work will have to demonstrate in the whole 'limits to growth' debate and the search for sustainable solutions that responding to the threat of global warming will pay more dividends than investing in other options. Given the challenges facing the human race in respect of population growth, poverty and disease, food production, energy generation, deforestation and overfishing, this is only likely to happen if climatological studies can make major new contributions to the solution of these problems. The uncertainties implicit in predicting climatic change reviewed here do not offer the prospect of this happening in the immediate future.

What will change the position is not any sudden leap forward in research, but a real change in the weather. In the light of incontrovertible evidence of a more fundamental shift in the climate, research into how we could adapt to this would be paramount. But a panic reaction to a few extreme events, which could be no more than a statistical blip, could do more harm than good. To take an example, it has been suggested that the 20-year cycle in US drought could lead to a massive drought around the end of the century across the Great Plains.[6] If this were to occur and proved to be significantly more extreme than the events of 1934 to 1936, it might be a trigger for reassessing thinking. To judge from the experience of 1988, long before three years was up and the exceptional nature of the event had been established, there would be huge pressure for political action. If this then proved premature, as the weather returned to more normal patterns, the damage to the credibility of those who had claimed that we had at last seen a sea-change would be all the greater. The fable of the boy crying 'wolf' comes to mind.

So the message is simple: if the weather is really going to matter, it has to matter all the time, and not just when it is selling newspapers. This means the climatological and meteorological communities have to resist the temptation to overexploit short-lived surges in interest. If they fail to do so, they cannot be surprised if the public hangs on to their every word during record-breaking weather, but lose interest when it returns to normal. Furthermore their flirtation with the limelight could have painful consequences. Their political paymasters could conclude later that not only have they failed to deliver what they promised, but once the hue and cry has died down, less funding is needed. This is why forecasters must, wherever possible, set quantifiable targets against which their progress and society's understanding of climatic change can be measured. This will enable decision-makers to adopt a measured approach to the 'precautionary principle', as any decisions on investing in schemes to prevent global warming must compete with other options for improving social well-being. By offering realistic guidance on what is known and what can be done, it will be possible to concentrate on action which can be taken promptly at comparatively low cost with the best prospects of avoiding more costly damage later, or of preventing irreversible effects which may result from action being delayed.

This demand for a clear statement of the 'skill' of any forecast may seem a bit hard on the meteorological and climatological communities. They can argue that they are doing their best. But, if forced to adopt a more rigorous approach, they may reap the benefit of calling the politicians' bluff. When confronted by a clear statement of the uncertainties in current analysis and the limited prospects of narrowing this down in the near future, governments will not be able to shelter behind disingenuous claims that what they are doing has been decided solely on the basis of scientific evidence. Instead they will have to spell out the political judgements that have been made in reaching decisions on whether or not to take action on the options for dealing with the social aspects of any current extreme weather and the threat of global warming. This more realistic approach may, however, be mutually beneficial to both sides if it serves to curb public expectations. In effect, the electorate will be told they will have to lump whatever the weather throws at them because, in the light of current scientific uncertainty about future climatic developments, governments have done all that can reasonably be expected given the competing demands for limited public funds. Only when it becomes clear that some real change is in progress could it be justified in increasing spending because in some ways the weather now matters more.

What researchers must not do is continually claim that if only they had more computing power, better observation networks and more sophisticated satellite technology, they would be able to come up with the answers. Worse still, if drawing up this 'wish list' implies that, by striving to stay at the cutting edge of science, the huge amount of painstaking work needed to make the fullest use of what has already been observed or discovered is all too much of a chore, policy-makers will become increasingly exasperated with requests for new and ever more expensive toys. The worst danger for researchers is to create the impression that, although fascinating, their goals are, like the end of the rainbow, ultimately unattainable. As a colleague once put it, on reading a detailed analysis of a field of work, 'just as I thought, more research is needed, but it won't give you the answer'.

8.6 Notes

1 Karl *et al.*(1996).
2 IPCC (1995), p. 168.
3 Yohe *et al.* (1996).
4 Cline (1992).
5 Dawkins (1982), p. 35.
6 Mitchell (1990).

References

Alley, R. B. *et al.* (1993). Abrupt increase in Greenland snow accumulation at the end of the Younger Dryas event. *Nature*, **362**, 527–9.

Baillie, M. G. L. (1995). *A Slice Through Time: Dendrochronology and Precision Dating*. Batsford.

Barnston, A. G. (1995). Our improving capability in ENSO forecasting. *Weather*, **50**, 419–30.

Beckerman, W. (1995). *Small is Stupid: Blowing the Whistle on the Greens*. Duckworth.

Beveridge, W. H. (1921). Weather and harvest cycles. *Economics Journal*, **31**, 429–47.

Bonan, G. B., Pollard, D. & Thompson, S. L. (1992). Effects of boreal forest vegetation on global climate. *Nature*, **359**, 716–18.

Borzenkova, I. I. & Zubakov, V. A. (1984). Climatic optimum of Holocene as a model of the global climate at the beginning of the 21st century. *Meteorol. i Gidrolog.*, N8, 69–77. Also *Sov. Met. Hydro.*, **8**, 52–8.

Bradley, R. S. & Jones, P. D. (eds) (1995). *Climate since AD 1500*. Routledge.

Briffa, K. R. *et al.* (1990). A 1400-year tree-ring record of summer temperatures in Fennoscandia. *Nature*, **346**, 434–39.

Briffa, K. R., Jones, P. D., Schweingruber, F. H , Shiyatov, S. G. & Cook, E. R. (1995). Unusual twentieth-century summer warmth in a 1000-year temperature record from Siberia. *Nature*, **376**, 156–9.

Broecker, W. S. (1994). Massive iceberg discharges as triggers for global climate change. *Nature*, **372**, 421–5.

Broecker, W. S. (1995a). Chaotic climate. *Scientific American*, **267**, No. 11, 44–50.

Broecker, W. S. (1995b). Cooling the tropics. *Nature*, **376**, 212–13.

Budyko, M. & Izrael, Yu. A. (eds) (1987). *Anthropogenic Climate Changes*. L. Gidrometeoizdat, 404 pp.

Burns, T. (1986). The interpretation and use of economic predictions. *Procedings of the Royal Society of London*, **A407**, 103–25.

Bryant, A. (1957). *The Turn of the Tide 1939–1943*. Collins.

Burroughs, W. J. (1982). The decline of walnut in English furniture-making. *Weather*, **37**, 272–4.

Burroughs, W. J. (1994). *Weather Cycles: Real or Imaginary?* Cambridge University Press.

Cane, M. A., Eshel, G. & Buckland, R. W. (1994). Forecasting Zimbabwean maize yield using eastern equatorial Pacific sea surface temperature. *Nature*, **370**, 2045.

Carpenter, R. (1966). *Discontinuity in Greek Civilisation*. Cambridge University Press.

Central Electricity Generating Board (1963). *Annual Report and Accounts 1962–3*. HMSO.

Cess, R. D. *et al.* (1991). Interpretation of snow–climate feedback as produced by 17 general circulation models. *Science*, **253**, 888–92.

Cess, R. D. *et al.* (1995). Absorption of solar radiation by clouds: observations versus models. *Science*, **267**, 496–9.

Chan, A., Savage, D., & Whittaker, R. (1995). A new Treasury model. *Government Economic Service Working Paper No. 128*. H.M. Treasury.

Chang, P., Ji, L. & Li, H. (1997). A decadal climate variation in the tropical Atlantic Ocean from thermodynamic air–sea interactions. *Nature*, **385**, 516–18.

Cline, W. R. (1992). *The Economics of Global Warming*. Institute for International Economics.

Crutchfield, J. P., Farmer, J. D., Packard, N. H. & Shaw, R. S. (1986). 'Chaos'. *Scientific American*, **256**, December, 38–49.

Currie, R. G. (1988). Lunar tides and the wealth of nations. *New Scientist*, **120**, 5 November, 52–5.

Dansgaard, W., Johnsen, S. J., Reeh, N., Gunderstrup, N., Clausen, H. B. & Hammer, C. U. (1975). Climatic changes, Norsemen and modern man. *Nature*, **255**, 24–8.

Dansgaard, W. *et al* (1993). Evidence of general instability of past climate from a 250-kyr ice-core record. *Nature*, **364**, 218–20.

Dawkins, R. (1982). *The Extended Phenotype*. Oxford University Press.

Eddy, J. A. (1976). The Maunder Minimum. *Science*, **192**, 1189–202.

Engelen, A. F. V. van, & Nellestijn, J.W. (1995). Monthly, seasonal, and annual means of the air temperature in tenths of centigrade in De Bilt, Netherlands, 1706–1995. KNMI.

Fink, A., Ulbrich, V. & Engel, H. (1996). Aspects of the January 1995 flood in Germany. *Weather*, **51**, 34–9.

Folland, C. K. & Parker, D. E. (1995). Correction of instrumental biases in historical sea surface temperature data. *Quarterly Journal Royal Meteorological Society*, **121**, 319–67.

Folland, C. K. & Rowell, D. E. (eds) (1995). Workshop of simulations of the climate of the twentieth century using GISST (28–30 November 1994). Hadley Centre CRTN 56, Bracknell, UK.

Gray, W. M. (1990). Strong association between West African rainfall and US landfall of intense hurricanes. *Science*, **249**, 1251–6.

Greenland Ice Core Project (GRIP) Members (1993). Climate instability during the last interglacial period recorded in the GRIP ice core. *Nature*, **364**, 203–7.

Grootes, P. M., Stuiver, M., White, J. W. C., Johnsen, S. & Jouzel, J. (1993). Comparison of oxygen isotope records from the GISP 2 and Greenland ice cores. *Nature*, **366**, 552–4.

Grove, J. M. (1988). *The Little Ice Age*. Methuen.

Haigh, J. D. (1996). The impact of solar variability on climate. *Science*, **272**, 981–4.

Hammer, C. U., Clausen, H. B. & Dansgaard (1980). Greenland ice sheet evidence of post-glacial volcanism and its climatic impact. *Nature*, **288**, 230–5.

Hannah, L. (1979). *Electricity Before Nationalisation*. Macmillan.

Hansen, J., Lacis, A., Ruedy, R. & Sato, M. (1992). Potential climate impact of Mount Pinatubo eruption. *Geophysical Research Letters*, **19**, 215–18.

Hansen, J. E., Wilson, H., Sato, M., Ruedy, R., Shah, K. P. & Hansen, E. (1995). Satellite and surface temperature data at odds? *Climate Change*, **30**, 103–17.

Hodell, D. A., Curtis, J. H. & Brenner, M. (1995). The role of climate in the collapse of classic Mayan civilization. *Nature*, **375**, 391–4.

Hoffman, J. S., Keyes, D. & Titus, J. G. (1983). *Projecting Future Sea Level Rise: Methodology, Estimates to the Year 2100, and Research Needs*. U.S. GPO 055–000–0236–3.GPO, Washington DC.

Hoskins, W. G. (1964). Harvest fluctuations and English economic history, 1480–1619. *The Agricultural History Review*, **12**, 28–46.

Hoskins, W. G. (1968). Harvest fluctuations and English economic history, 1620–1659. *The Agricultural History Review*, **16**, 15–31.

Hulme, M. & Jones, P. D. (1991). Temperatures and windiness over the United Kingdom during the winters of 1988/89 and 1989/90 compared with previous years. *Weather*, **46**, 126–36.

Hulme, M. *et al.* (1992). Seasonal rainfall forecasting for Africa. Part 1. *International Journal of Environmental Studies*, **39**, 245–56, and Part II. *International Journal of Environmental Studies*, **40**, 103–21.

Imbrie, J. & Imbrie, J. Z. (1980). Modelling the climatic response of orbital variations. *Science*, **207**, 943–53.

IPCC (1990). *Climate Change: The IPCC Scientific Assessment*. J. T. Houghton, G. J. Jenkins & G. G. Ephraums (eds). Cambridge University Press, UK, 365pp.

IPCC (1992). *Climate Change 1992: the Supplementary Report to IPCC Scientific Assessment*, J. T. Houghton, B. A. Callander & S. K. Varney (eds). Cambridge University Press, UK.

IPCC (1994). *Climate Change 1994: Radiative Forcing of Climate and an Evaluation of the IPCC IS92 Emission Scenarios*, J. T. Houghton, L. G. Meira Filho, J. Bruce, Hoesung Lee, B. A.

Callendar, E. Haites, N. Harris, & K. Maskell. (eds). Cambridge University Press, UK.

IPCC (1995). *Climate Change 1995: The Science of Climate Change.* J. T. Houghton, L. G. Meira Filho, B. A. Callendar, N. Harris, A. Kattenberg & K. Maskell. (eds). Cambridge University Press, UK.

Kaiser, R. G. (1974). *Cold Winter, Cold War.* Weidenfeld & Nicholson.

Karl, T. R., Knight, R. W., Easterling, D. R. & Quayle, R. G. (1996). Indices of climate change for the United States. *Bulletin of the American Meterological Society*, 77, 279–92.

Kates, R. W., Ausubel, J. H. & Berberain, M. (eds) (1985). *Climate Impact Assessment: Studies of the Interaction of Climate and Society.* Wiley.

Kennedy, H. (1986). *The Prophet and the Age of the Caliphates: The Islamic Near East from the 6ᵗʰ to 11ᵗʰ Centuries.* Longmans.

Kerr, R A. (1995). Darker clouds promise brighter future for climate models. *Science*, 267, 454.

Kripalani, R. H. & Kulkarni, A. (1997). Climatic impact of the El Niño/La Niña on the Indian Monsoon: a new perspective. *Weather*, 52, 39–46.

Krishna Kumar, K., Soman, M. K. & Rupa Kuma, K. (1995). Seasonal forecasting of Indian summer monsoon rainfall: A review. *Weather*, 50, 449–67

Laird, K. R., Fritz, S. C., Maasch, K. A. & Cumming B. F. (1996). Greater drought intensity and frequency before AD 1200 in the Northern Great Plains, USA. *Nature*, 384, 552–54.

Lamb, H. H. (1972). *Climate: Present, Past and Future.* Volume 1. Methuen.

Lamb, H. H. (1995). *Climate, History and the Modern World* (2nd edn). Routledge.

Landsea, C. W. (1993). A climatology of intense (or major) Atlantic hurricanes. *Monthly Weather Review*, 121, 1703–13.

Landsea, C. W., Gray, W. M., Mielke, Jr. P. W. & Berry, J. K. (1994). Seasonal forecasting of Atlantic hurricane activity. *Weather*, 49, 273–284.

Lawrence, E. N. (1972). The earliest known journal of the weather. *Weather*, **27**, 494–501.

Le Roy Ladurie, E. (1972). *Times of Feast, Times of Famine: A History of Climate Since the Year 1000*. George Allen & Unwin.

Le Roy Ladurie, E. & Baulant, M. (1980). Grape harvests from the fifteenth through the nineteenth centuries. *Journal of Interdisciplinary History*, **10**, 839–49.

Li, X, Maring, H., Savoie, D., Voss, K. & Prospero, J. M. (1996). Dominance of mineral dust in aerosol light scattering in the North Atlantic trade winds. *Nature*, **380**, 416–19.

Liddell Hart, B. H. (1970). *History of the Second World War*. Cassell.

Lott, J. N. (1994). The US summer of 1993: a sharp contrast in weather extremes. *Weather*, **49**, 370–83.

Lorenz, E. N. (1982). Atmospheric predictability experiments with a large numerical model. *Tellus*, **34**, 505–13.

Lorenz, E. N. (1993). *The Essence of Chaos*. UCL Press.

Lucas, H. S. (1930). The Great European Famine of 1315, 1316 and 1317, *Speculum*, **V**, reprinted in Carus-Wilson, E. M. (ed.) (1962). *Essays in Economic History*, Vol. II, 49–72. Arnold.

Mango, C. (1988). *Byzantium: The Empire of the New Rome*. Weidenfeld & Nicholson.

Manley, G. (1974). Central England Temperatures: monthly means 1659 to 1973. *Quarterly Journal Royal Meteorological Society*, **100**, 389–405.

Markson, R. (1978). Solar modification of atmospheric electrification and possible implications for the Sun-weather relationship. *Nature*, **244**, 197–200.

Maunder, W. J. (1970). *The Value of the Weather*. Methuen.

Maunder, W. J. (1987). *The Uncertainty Business: Risks, Implications & Responses*. Methuen.

McCallum, E. (1990). The Burns Day Storm, 25 January 1990. *Weather*, **45**, 166–73.

Meteorological Office (1987). *The Storm of 15/16 Oct 1987*. UK Meteorological Office.

Mintzer, I. M. (ed) (1992). *Confronting Climate Change: Risks, Implications and Responses*. Cambridge University Press.

Mitchell, J. F. B., Johns, T. C., Gregory, J. M. & Tett, S. F. B. (1995). Climate response to increasing levels of greenhouse gases and sulphate aerosols. *Nature*, 376, 501–4.

Mitchell, J. M., Stockton, C. W. & Meko, D. M. (1979). Evidence of a 22-year rhythm of drought in the Western United States related to the Hale Solar Cycle since the 17th century. In *Solar-Terrestrial Influences on Weather and Climate*, ed. B. M. McCormac & T. A. Seliga. D. Reidel Publishing Co.

Mitchell, J. M. (1990) Climatic variability: past, present & future. *Climatic Change*, 16, 231–46.

Montieth, J. L. (1981). Climatic variation and the growth of crops. *Quarterly Journal Royal Meteorological Society*. 107, 749–74.

Neumann, J. (1992). The severe winter 1941/42 in the European Soviet Union: frost casualties among the troops on the Eastern Front, *Weather*, 47, 480–83.

Nordhaus, W. D. (1990). To slow or not to slow: the economics of the greenhouse effect. *The Economics Journal*, 101, 920–27.

Nordhaus, W. D. (1992). An optimal transition path for controlling greenhouse gases. *Science*, 258, 135–9.

Nutter, F. W. (1994). The role of government in the United States in addressing natural catastrophes and environmental exposures. *The Geneva Papers on Risk and Insurance*. No. 72, 244–56.

Oliver, J. E. (1981). *Climatology: Selected Applications*. Edward Arnold.

Ormerod, P. (1994). *The Death of Economics*. Faber & Faber.

Palmer, T. (1989). A weather eye on unpredictability. *New Scientist*, 124, 11 November, 56–9.

Palmer, T. (1993). A nonlinear dynamical perspective on climate change. *Weather*, 48, 314–25.

Parker, D. E., Legg, T. P. & Folland, C. K. (1992). A new daily Central England temperature series, 1772–1991. *International Journal of Climatology*. 12, 317–42.

Parry, M. & Duncan, R. (eds) (1995). *The Economic Implications of Climate Change in Britain*. Earthscan.

Pfister, C. (1995). Monthly temperature and precipitation in central Europe 1525–1979: quantifying documentary evidence on weather and its effects. In *Climate Since AD 1500*, ed. R. S. Bradley & P. D. Jones, [Chapter 6], Routledge. [Data on temperature and preciptation indices available on disk from the National Geophysical Data Center, Boulder, Co 80309, USA.]

Phelps-Brown, E. H. & Hopkins, S. V. (1956). Seven centuries of the prices of consumables, compared with builders' wage-rates. *Economica*, **xxiii**, 296–314. (Reprinted in Phelps-Brown & Hopkins (1981).

Phelps-Brown, E. H. & Hopkins, S. V. (1981). *A Perspective on Wages and Prices*. Methuen.

Philander, S. G. H. (1983). El Niño Southern Oscillation. *Nature*, **302**, 295–301.

Pimental, D. *et al.* (1995). Environmental and economic costs of soil erosion and conservation benefits. *Science*, **267**, 1117–23.

Rampino, M. R. & Self, S. (1992). Volcanic winter and accelerated glaciation following the Toba super-eruption. *Nature*, **359**, 50–2.

Rosenzweig, C. (1994). Maize suffers a sea-change. *Nature*, **370**, 174–6.

Rosenzweig, C. & Parry, M. L. (1994). Potential impact of climate change on world food supply. *Nature*, **367**, 133–8.

Sandars, N. K. (1985). *The Sea Peoples: Warriors of the Ancient Mediterranean*. Thames & Hudson.

Schmidlin, T. W. (1993). Impacts of severe winter weather during December 1989 in the Lake Erie snowbelt. *Journal of Climate*, **6**, 759–67.

Sheratt, A. (1980). *Cambridge Encyclopedia of Archaeology*. Cambridge University Press, UK.

Solomou, S. (1990). *Phases of Economic Growth 1850–1973: Kondratieff waves and Kuznets swings*. Cambridge University Press, UK.

Spencer, R. W. & Christy, J. R. (1990). Precise monitoring of global temperature trends from satellites. *Science*, **247**, 1558–62.

Stolfi, R. H. S. (1980). Chance in history: the Russian winter of 1941–1942. *History*, **65**, 214–28.

Stommel, H. & Stommel, E. (1979). The year without a summer. *Scientific American*, **240**, June, 134–40.

Stone, R. (1995). If the mercury soars, so may health hazards. *Science*, **267**, 957–8.

Stothers, R. B. (1984). Mystery cloud of AD 536. *Nature*, **307**, 344–5.

Stouffer, R. J., Manabe, S. & Vinnilcov, K. Ya. (1994). Model assessment of the role of natural variability in recent global warming. *Nature*, **367**, 634–6.

Strzepek, K. M. & Smith, J. B. (eds) (1995). *As Climate Changes: International Impacts and Implications*. Cambridge University Press, UK.

Taylor, K. C. *et al.* (1993). The 'flickering switch' of late Pleistocene climate change. *Nature*, **361**, 432–6.

Tengen, I., Lacis, A. A. & Fung, I. (1996). The influence on climate forcing of mineral aerosols from disturbed soils. *Nature*, **380**, 419–22.

Tennekes, H. (1992). Karl Popper and the accountability of numerical weather forecasting. *Weather*, **47**, 343–6.

Thomas, D. S. G. & Middleton, N. J. (1994). *Desertification: Exploding the Myth*. Wiley.

Thompson, J. E. S. (1993). *The Rise and Fall of Maya Civilisation*. Plimlico.

Tinsley, B. A. (1988). The solar cycle and the QBO influences on the latitude of storm tracks in the North Atlantic. *Geophysical Research Letters*, **15**, 409–12.

Titow, J. (1960). Evidence of weather in the account rolls of the bishopric of Winchester 1209–1350. *Economic History Review*, **12**, 360–407.

Toumi, R., Bekki, S. & Law, K. (1995). Indirect influence of ozone depletion on climate forcing by clouds. *Nature*, **372**, 348–51.

Trenberth, K. E. (ed.) (1992). *Climate System Modelling*. Cambridge University Press.

Tuchman, B. W. (1978). *A Distant Mirror: The Calamitous 14th Century*. Macmillan.

Tucker, C. J., Dregne, H. E. & Newcomb, W. W. (1991). Expansion and contraction of the Sahara Desert from 1980 to 1990. *Science*, **253**, 299–301.

Tylecote, A. (1991). *The Long Wave in the World Economy: The Current Crisis in Historical Perspective*. Routledge.

Van den Dool, H. M., Krijnen, H. J. & Schuurmans, C. J. E. (1978).

Average winter temperaures at De Bilt (Netherlands): 1634–1977. *Climatic Change*, 1, 319–30.

Weaver, A. J., & Hughes, T. M. C. (1994). Rapid interglacial climate fluctuations driven by North Atlantic ocean circulation. *Nature*, 367, 447–50.

Weaver, A. J., Sarachik, E. S. & Marotzke, J. (1991). Freshwater flux forcing of decadal and interdecadal oceanic variability. *Nature*, 353, 836–8.

Wigley, T. M. L. & Atkinson, T. C. (1977). Dry years in south-east England since 1698. *Nature*, 265, 431–4.

Wigley, T. M. L., Lough, J. M. & Jones, P. D. (1984). Spatial patterns of precipitation in England and Wales and a revised homogenous England and Wales precipitation series. *Journal of Climatology*, 4, 1–25.

Wigley, T. M. L., Ingram, M. J. & Farmer, G. (1981). *Climate and History: Studies in Past Climates and their Impact on Man.* Cambridge University Press, UK.

Wiscombe, W. J. (1995). An absorbing mystery. *Nature*, 376, 466–7.

World Meteorological Organisation (WMO) (1995). *The Global Climate System Review: Climate System Monitoring June 1991–November 1993.* WMO.

Wrigley, E. A. & Schofield, R. S. (1989). *The Population History of England 1541–1871: A reconstruction.* Cambridge University Press, UK.

Yohe, G., Neumann, J., Marshall, P. & Ameden, P. (1996). The economic cost of greenhouse-induced sea-level rise for developed property in the United States. *Climate Change*, 32, 387–410.

Acknowledgements

Because this book is the distillation of lengthy personal experience, it is difficult to identify all the people who have helped me to form a view on the many facets of extreme weather, climate change and their impact on society. Among the meteorological community I would like to thank Chris Folland, David Parker, Tony Slingo, John Mitchell, Bruce Callendar, Tim Palmer, Tony Hollingsworth, Anders Persson, Austen Woods, Phil Jones, Tom Holt, Aryan van Engelen, Tom Karl and John Christy for helpful discussions, and the provision of data and other material essential for the completion of this book. I am also grateful to Ed Bartholomew, Mark Lynch, Nancy Sandars, Mary Spence and Andrew Writing for help in locating pictures. More generally I am indebted to my many friends and colleagues in the UK Departments of Energy and Health, with whom I worked between 1974 and 1995, and whose stimulating and varied personal views on economics, politics, and the challenges of developing policy on wide-ranging multidisciplinary social issues had such an influence on my thinking. I would like also to thank Janice Robertson for her help in getting the text into a tidy form. Finally, I am deeply indebted to my wife who, as always, helped and supported me throughout the lengthy gestation of this book.

Index

Printed in the United Kingdom
by Lightning Source UK Ltd.
108646UKS00001B/106-108